Chapter 1: Introduction to Nicotinic Acetylcholine Receptors (nAChRs)

History of Acetylcholine Discovery

The history of acetylcholine (ACh) dates back to the early 20th century, marking a groundbreaking discovery in the field of neuroscience. Acetylcholine was first identified as a neurotransmitter in 1921 by Sir Henry Hallett Dale, an English pharmacologist. Dale, through his experiments, demonstrated that acetylcholine played a crucial role in the transmission of nerve impulses in both the central and peripheral nervous systems. For this discovery, Dale was awarded the Nobel Prize in Physiology or Medicine in 1936, a recognition of the significant implications of his work.

The functional characterization of acetylcholine continued with further contributions from Otto Loewi, who elucidated the role of acetylcholine in the transmission of signals across synapses. Loewi's famous experiment in 1921, which demonstrated the chemical nature of nerve signaling, laid the foundation for understanding neurotransmission and acetylcholine's pivotal role in that process.

Since then, extensive research into acetylcholine has revealed that it is involved not only in muscle contraction but also in regulating numerous physiological processes, including heart rate, digestion, and cognitive function. One of the most significant advances was the identification of the specific receptors to which acetylcholine binds: the nicotinic and muscarinic acetylcholine receptors, both of which are essential for ACh's diverse biological effects.

Role of Acetylcholine in the Nervous System

Acetylcholine is a key neurotransmitter that facilitates communication between neurons and between neurons and other cells, such as muscle cells. It is synthesized in cholinergic neurons and is released at synapses, where it binds to receptor proteins on the postsynaptic cell, triggering a range of cellular responses.

In the peripheral nervous system (PNS), acetylcholine plays a pivotal role in muscle contraction and the function of the autonomic nervous system, which controls involuntary bodily functions such as heart rate, blood pressure, and digestion. In the central nervous system (CNS), acetylcholine is crucial for cognitive processes, including learning, attention, and memory. It has also been implicated in the modulation of sleep-wake cycles and in regulating mood and emotional responses.

Given its widespread influence on both voluntary and involuntary bodily functions, understanding acetylcholine's interaction with its receptors is fundamental to advancing therapeutic strategies for a range of disorders, from neurodegenerative diseases like Alzheimer's to addiction and muscle-related disorders.

Overview of Receptor Types: Muscarinic vs. Nicotinic

Acetylcholine exerts its effects by binding to two main types of receptors: muscarinic acetylcholine receptors (mAChRs) and nicotinic acetylcholine receptors (nAChRs). While both are involved in neurotransmission, they differ significantly in their structure, mechanism of action, and physiological effects.

Muscarinic Acetylcholine Receptors (mAChRs)

These receptors are G-protein coupled receptors (GPCRs) and mediate slower, longer-lasting effects. They are predominantly found in the heart, smooth muscles, and glands, where they regulate functions such as heart rate and secretion. There are five subtypes of muscarinic receptors (M1–M5), each playing distinct roles in various tissues.

Nicotinic Acetylcholine Receptors (nAChRs)

Unlike muscarinic receptors, nicotinic receptors are ligand-gated ion channels that mediate fast synaptic transmission by allowing the flow of ions, such as sodium (Na^+) and calcium (Ca^{2+}), across the cell membrane. These receptors are found in both the central and peripheral nervous systems, including at neuromuscular junctions and autonomic ganglia. nAChRs are particularly important in mediating the rapid transmission of signals across synapses and are involved in processes such as muscle contraction and neuronal signaling.

While mAChRs and nAChRs share a common ligand—acetylcholine—their distinct structures and mechanisms of action result in different physiological outcomes. Nicotinic receptors, due to their role in fast synaptic transmission, have a more immediate and short-lived effect compared to the slower, more sustained actions of muscarinic receptors.

Structure and Function of nAChRs

Nicotinic acetylcholine receptors (nAChRs) are integral membrane proteins that act as ion channels, opening in response to acetylcholine binding. These receptors are composed of five subunits that form a pentameric structure around a central ion channel. Each subunit is made up of four transmembrane helices, and they are organized around a central pore that allows the passage of ions when activated.

Subunit Composition

The subunit composition of nAChRs can vary, leading to different receptor subtypes with distinct properties. The most common subtypes are composed of two α (alpha) subunits and three other subunits, which can be β (beta), γ (gamma), δ (delta), or ε (epsilon) in muscle-type receptors (located at neuromuscular junctions) or α and β subunits in neuronal receptors (found in the brain and autonomic ganglia). The α subunits are the primary binding sites for acetylcholine, and their interaction with ACh leads to receptor activation.

Ion Channel Function

Upon acetylcholine binding, the nAChR undergoes a conformational change, which opens the ion channel. This allows the influx of cations, primarily sodium (Na^+) and calcium (Ca^{2+}), and the efflux of potassium (K^+) ions. This ion movement depolarizes the postsynaptic membrane, generating an excitatory postsynaptic potential (EPSP) and propagating the signal across the synapse.

Desensitization and Adaptation

Nicotinic receptors are subject to a process known as desensitization. Prolonged or continuous exposure to acetylcholine can lead to a decrease in receptor responsiveness, even if the neurotransmitter is still present. This desensitization serves as a protective mechanism to prevent overstimulation of the postsynaptic cell. Receptors may also undergo adaptation, adjusting their sensitivity in response to changes in the environment or receptor activity.

Tissue Localization

nAChRs are found in a variety of tissues, with distinct functional roles in each. In the peripheral nervous system, they are critical for neuromuscular transmission, where their activation leads to muscle contraction. In the central nervous system, they are involved in regulating synaptic plasticity, cognition, and attention. Their role in the brain is particularly significant in the regulation of neurotransmitter release, influencing dopamine, serotonin, and glutamate systems, which are important for mood and learning processes.

In summary, nicotinic acetylcholine receptors are central to the rapid transmission of signals in both the central and peripheral nervous systems. They facilitate critical functions such as muscle contraction and synaptic plasticity, making them essential targets for drug development in a range of therapeutic areas, from neurology to addiction treatment. Understanding the structure and function of these receptors provides the foundation for exploring their role in health and disease, which will be explored further in the chapters that follow.

Summary

This introductory chapter provided an overview of the nicotinic acetylcholine receptors (nAChRs), focusing on their discovery, biological significance, and structural characteristics. As we continue, the following chapters will delve deeper into the molecular biology, biochemistry, and pharmacology of nAChRs, exploring how these receptors influence various physiological processes and their involvement in neurological disorders, addiction, and disease mechanisms. The goal is to equip readers with a comprehensive understanding of nAChRs, laying the groundwork for mastering their role in both basic and clinical sciences.

Chapter 2: Molecular Biology of Nicotinic Acetylcholine Receptors (nAChRs)

Genetic Encoding of nAChRs

The nicotinic acetylcholine receptor (nAChR) is a critical protein in the body, involved in both the central and peripheral nervous systems. These receptors are encoded by a family of genes that produce distinct subunits, each contributing to the overall structure and function of the receptor. The process of genetic encoding for nAChRs begins with the transcription of the corresponding genes, followed by translation to produce polypeptides that eventually assemble into functional receptors.

There are multiple genes that encode for nAChR subunits. In humans, the gene family responsible for the synthesis of nAChRs consists of 16 genes, including α (alpha), β (beta), γ (gamma), δ (delta), and ε (epsilon) subunits. Each subunit is encoded by a separate gene, and the composition of these subunits determines the specific properties and functions of the receptor. The various combinations of subunits lead to a diverse range of receptor subtypes, each of which has unique functional properties, such as ion channel conductance and pharmacological sensitivity.

The genetic encoding process begins in the cell nucleus, where specific genes are transcribed into mRNA. This mRNA then undergoes translation in the cytoplasm to form protein subunits. The assembled subunits then undergo various post-translational modifications to mature into functional receptors.

Gene Families Involved in nAChR Synthesis

The primary gene families involved in the synthesis of nAChRs include the **CHRNA** (cholinergic receptor, nicotinic, alpha), **CHRNB** (cholinergic receptor, nicotinic, beta), **CHRNG** (cholinergic receptor, nicotinic, gamma), **CHRND** (cholinergic receptor, nicotinic, delta), and **CHRNAE** (cholinergic receptor, nicotinic, epsilon). These gene families encode the different subunits that come together to form the functional nAChR.

Alpha Subunits (α)

The α subunits (α2–α10) are the most critical components in terms of acetylcholine binding. They each contribute to the receptor's activation and conductance properties. For example, the α1 subunit is primarily found in muscle-type receptors at the neuromuscular junction, whereas α4 and α7 subunits are more prevalent in neuronal receptors, such as those in the brain.

Beta Subunits (β)

The β subunits (β2–β4) typically associate with α subunits to form functional receptors. The β subunits play a key role in determining the overall ion conductance and the receptor's pharmacological properties.

Gamma (γ) and Epsilon (ε) Subunits

In muscle-type receptors, the γ subunit is present during early development and transitions to the ε subunit in adult tissue. The transition from γ to ε is important for proper receptor functionality and is essential for neuromuscular transmission.

Delta (δ) Subunit

The δ subunit is always present in muscle-type receptors and contributes to the formation of the ion channel, participating in the overall structure of the receptor.

Together, these subunits combine to form different nAChR subtypes, with distinct anatomical distributions and functional characteristics.

mRNA and Protein Expression of nAChRs

Once the genes encoding the nAChR subunits are transcribed into mRNA, they are translated into protein subunits in the cytoplasm. These subunits then undergo several folding and modification steps before assembling into the mature receptor complexes.

The expression of nAChR subunits is tightly regulated at both the transcriptional and post-transcriptional levels. In many tissues, the synthesis of nAChR subunits can be influenced by factors such as neuronal activity, age, and disease states. For example, in response to chronic nicotine exposure, the expression of certain nAChR subunits, particularly those in the brain, can increase as a part of a compensatory mechanism. This change in gene expression can contribute to the development of tolerance and addiction.

In the central nervous system, nAChR expression is highly tissue-specific, with certain receptor subtypes being enriched in specific regions of the brain. For example, the α4β2 subtype is predominant in areas involved in cognition, such as the hippocampus, while the α7 subtype is found in regions related to sensory processing and neuroinflammation.

In the peripheral nervous system, the expression of nAChRs is also highly regulated. For example, the neuromuscular junction primarily expresses α1β1δγ subunit combinations, whereas the autonomic ganglia have different compositions, including α3β4 subunits.

Localization in Different Tissues

The localization of nAChRs varies significantly across different tissues, and this diversity of receptor expression plays a key role in the physiological functions of acetylcholine in both the central and peripheral nervous systems.

Central Nervous System (CNS)

In the CNS, nAChRs are widely distributed, with high densities in areas responsible for cognitive functions such as the hippocampus, cerebral cortex, and basal ganglia. The α4β2 and α7 subtypes are the most common, and they play a key role in modulating neurotransmitter release, influencing cognitive processes such as learning, attention, and memory. Additionally, nAChRs in the brain are involved in the regulation of other neurotransmitter systems, including dopamine, serotonin, and glutamate, which have broad effects on mood, motivation, and neural plasticity.

Peripheral Nervous System (PNS)

In the PNS, nAChRs are found at neuromuscular junctions, where they mediate the communication between motor neurons and muscle cells. The α1β1δγ/ε subunit combinations are critical for this function. The activation of these receptors leads to the opening of ion channels, resulting in muscle contraction. nAChRs are also found in autonomic ganglia, where they mediate the transmission of signals between pre- and post-ganglionic neurons, thus contributing to the regulation of autonomic functions such as heart rate and digestion.

Other Tissues

Recent research has shown that nAChRs are also expressed in non-neuronal tissues such as immune cells, epithelial cells, and even in the endothelium of blood vessels. In immune cells, nAChRs are involved in the modulation of inflammation, and their activation can influence the immune response. In epithelial cells, nAChRs play a role in cell signaling, while their presence in endothelial cells suggests a role in regulating vascular tone and blood flow.

Developmental Changes in nAChR Localization

The expression pattern of nAChRs also changes during development. For instance, during fetal development, the γ subunit is present in muscle-type receptors at neuromuscular junctions, but in adulthood, this is replaced by the ε subunit. Similarly, the expression of nAChRs in the brain undergoes changes as the nervous system matures, with certain receptor subtypes becoming more prevalent in specific brain regions.

In summary, the molecular biology of nAChRs is complex, involving a variety of subunits encoded by distinct genes, which then combine to form functional receptors. The expression of these receptors is tightly regulated and varies between different tissues and developmental stages. As we continue to explore nAChRs in the following chapters, understanding their molecular and genetic foundations will provide insight into their diverse physiological roles and potential therapeutic applications.

Summary

Chapter 2 has provided an overview of the molecular biology of nicotinic acetylcholine receptors (nAChRs), focusing on their genetic encoding, the gene families involved in nAChR synthesis, mRNA and protein expression, and their tissue-specific localization. Understanding these molecular mechanisms is crucial for appreciating the diversity and functional significance of nAChRs across different tissues and systems in the body. With this foundational knowledge, we can now explore the biochemical properties of nAChRs in greater detail in the next chapter.

Chapter 3: Biochemistry of Nicotinic Acetylcholine Receptors (nAChRs)

The nicotinic acetylcholine receptor (nAChR) is a highly sophisticated and dynamic protein complex, central to both the transmission of nerve impulses and a variety of physiological functions. Understanding its biochemistry is crucial for understanding how these receptors mediate rapid and precise signaling within the nervous system. This chapter will delve into the key biochemical features of nAChRs, including receptor subtypes and their significance, ligand binding, ion channel function, and the mechanisms of desensitization and adaptation.

Receptor Subtypes and Their Significance

Nicotinic acetylcholine receptors are classified into several subtypes based on their subunit composition, anatomical localization, and functional roles. The most well-studied nAChRs are the ones found in the central and peripheral nervous systems, and at neuromuscular junctions. The variety in subunit combinations determines the functional properties and pharmacological characteristics of each receptor subtype. Understanding these variations is essential for uncovering the complexity of nAChR signaling.

Muscle-Type nAChRs (α1β1δγ/ε)

These receptors are predominantly found at the neuromuscular junction, where they mediate the transmission of nerve signals to muscle fibers, leading to muscle contraction. The classic composition of muscle-type nAChRs includes two α1 subunits, one β1, one δ, and one γ (in fetal tissue) or ε (in adult tissue) subunit. The transition from the γ to the ε subunit in postnatal development is crucial for the proper function of these receptors in adult muscle.

Neuronal-Type nAChRs (α2–α10, β2–β4)

These receptors are primarily located in the central nervous system and autonomic ganglia. The subtypes within this group differ in their α and β subunit compositions, with common combinations including α4β2, α3β4, and α7. The α4β2 subtype is particularly prevalent in the brain and is involved in processes such as learning, memory, and attention, whereas the α7 subtype is important for fast neurotransmitter release and plays a role in inflammatory processes.

Subunit Variability

The diversity of nAChR subtypes results from the combination of different α, β, δ, γ, and ε subunits. This allows nAChRs to adapt to a wide variety of physiological contexts, making them a target for drug development in treating neurological and muscular disorders, as well as addiction. For instance, the α7 subtype is highly sensitive to nicotine and is implicated in the modulation of synaptic plasticity and neuroinflammation, making it an attractive target for research into Alzheimer's disease and other neurodegenerative disorders.

Ligand Binding and Receptor Activation

The function of nAChRs is largely dictated by how they interact with ligands, particularly acetylcholine (ACh), and the subsequent activation of the ion channel. Ligand binding is a highly specific process that occurs at the extracellular domain of the receptor, and it triggers a conformational change that opens the ion channel.

Acetylcholine Binding

Acetylcholine, the endogenous ligand, binds to specific sites on the α subunits of the receptor. Each nAChR subunit has an extracellular domain that contains a binding site for acetylcholine. For activation, acetylcholine molecules bind to the α subunits, causing a conformational change that opens the central ion channel. This process is highly efficient and allows for rapid synaptic transmission.

Ion Channel Opening

Once acetylcholine binds to the receptor, the ion channel undergoes a conformational shift that allows the passage of cations, primarily sodium (Na^+), potassium (K^+), and calcium (Ca^{2+}). This influx of ions depolarizes the postsynaptic membrane, generating an excitatory postsynaptic potential (EPSP) that propagates the signal to the next neuron or muscle cell. The ion conductance of nAChRs is typically fast and short-lived, facilitating rapid signaling in both neuronal and muscle cells.

Pharmacological Ligands

In addition to acetylcholine, nAChRs can be activated by various other ligands, both synthetic and natural. Nicotine is one of the most well-known exogenous ligands, binding with high affinity to several nAChR subtypes, particularly those involving the α4β2 and α7 combinations. Other compounds, such as certain pharmaceuticals and toxins, can also bind to these receptors, either activating them (agonists) or blocking them (antagonists), leading to a range of effects on the nervous system.

Ion Channel Function

Once activated, nAChRs function as ion channels that allow the flow of ions across the cell membrane. The ion channel is composed of a central pore formed by the subunits of the receptor. The opening of this channel is essential for the receptor's role in synaptic transmission and muscle contraction.

Ion Selectivity

nAChRs are non-selective cation channels, meaning they allow the passage of a variety of cations, including Na^+, K^+, and Ca^{2+}. The influx of sodium ions leads to depolarization of the postsynaptic membrane, while the efflux of potassium ions contributes to repolarization. Calcium ions, although less abundant in the current, play a role in signal transduction, particularly in neuronal signaling and synaptic plasticity.

Kinetics of Channel Opening

The opening and closing of the nAChR ion channel occurs rapidly, allowing for fast synaptic transmission. The duration of the channel's opening is very brief, typically in the order of milliseconds, before it closes again. This rapid on/off mechanism is essential for nAChRs to participate in the fast-paced communication required by the nervous system.

Excitatory Synaptic Transmission

As a result of their ion conductance properties, nAChRs mediate excitatory synaptic transmission, meaning that they increase the likelihood of the postsynaptic cell firing an action potential. This is particularly important in neurons, where nAChRs facilitate communication between neurons and between neurons and other types of cells (e.g., muscle cells at neuromuscular junctions).

Desensitization and Adaptation Mechanisms

One of the unique features of nAChRs is their ability to undergo desensitization and adaptation. These processes ensure that the receptors do not remain continuously activated in the presence of acetylcholine or other ligands, which would lead to overstimulation of the postsynaptic cell.

Desensitization

Desensitization is a process by which, after an initial activation, the receptor becomes unresponsive to further ligand binding despite the presence of the ligand. This occurs when the receptor undergoes conformational changes that prevent the ion channel from opening, even though acetylcholine is still bound to the receptor. Desensitization protects the postsynaptic cell from prolonged depolarization and potential damage due to excessive ion influx. For example, nicotine, when present for extended periods, leads to desensitization of nAChRs, a process thought to contribute to the development of nicotine tolerance.

Receptor Adaptation

Adaptation is a more gradual process where the expression or sensitivity of nAChRs can change in response to chronic changes in ligand exposure. For example, prolonged exposure to nicotine can lead to an increase in nAChR expression in certain brain regions, contributing to tolerance and dependence. Adaptation mechanisms are essential for maintaining homeostasis and adjusting the receptor response to changes in neurotransmitter levels.

Reversal of Desensitization

Interestingly, nAChRs are capable of "reversing" desensitization when the ligand is removed or the receptor is reset. This adaptability is crucial for the proper function of nAChRs in dynamic environments, allowing the receptor to return to a responsive state after ligand-induced desensitization.

Conclusion

The biochemistry of nicotinic acetylcholine receptors is central to their ability to mediate rapid synaptic transmission and influence various physiological processes. Through their diverse subtypes, nAChRs participate in numerous functions, including cognitive processes, muscle contraction, and neuroinflammation. Their ability to bind ligands such as acetylcholine and nicotine, conduct ions, and undergo desensitization and adaptation makes them crucial for maintaining the balance of neural activity. Understanding the biochemical properties of these receptors provides valuable insights into their role in health and disease and informs the development of therapeutic interventions targeting nAChRs.

Summary

This chapter has outlined the biochemistry of nicotinic acetylcholine receptors (nAChRs), covering key aspects such as receptor subtypes, ligand binding, ion channel function, and desensitization mechanisms. Understanding these processes is fundamental for mastering how nAChRs operate within the nervous system and how they contribute to various physiological functions. In the next chapter, we will explore the physiological roles of nAChRs in synaptic transmission and their involvement in both central and peripheral nervous systems.

Chapter 4: Physiology of Nicotinic Acetylcholine Receptors (nAChRs)

The nicotinic acetylcholine receptor (nAChR) is essential to the functioning of the nervous system, serving as the gateway through which acetylcholine (ACh) transmits signals between nerve cells and between nerve and muscle cells. The physiological roles of nAChRs are diverse and far-reaching, impacting everything from basic synaptic transmission to more complex processes such as motor function and cognitive performance. This chapter will explore the role of nAChRs in synaptic transmission, their functions within both the central and peripheral nervous systems, their connection to motor function, and their interaction with other neurotransmitter systems.

Role of nAChRs in Synaptic Transmission

At its core, the function of nAChRs is to mediate synaptic transmission. This process is vital for communication between neurons and between neurons and other types of cells, including muscle cells. Synaptic transmission occurs when a neurotransmitter, such as acetylcholine, is released from the presynaptic neuron and binds to receptors on the postsynaptic cell, leading to an excitation or inhibition of that cell.

Mechanism of Action

In cholinergic synapses, acetylcholine is released from vesicles in response to an action potential reaching the presynaptic terminal. The acetylcholine binds to the nicotinic receptors located on the postsynaptic membrane. Upon binding, the nAChR undergoes a conformational change, opening the ion channel and allowing the passage of ions, primarily sodium (Na^+), potassium (K^+), and calcium (Ca^{2+}). This ion flow depolarizes the postsynaptic membrane, initiating an action potential in the postsynaptic cell if the depolarization is sufficient to reach the threshold for firing.

Excitatory Postsynaptic Potential (EPSP)

The influx of sodium and calcium ions into the postsynaptic neuron generates an excitatory postsynaptic potential (EPSP), increasing the likelihood of that neuron firing an action potential. This rapid transmission of electrical signals across the synapse is a hallmark of nAChRs' role in synaptic communication.

Fast and Short-Lived Response

The nAChR's fast, ligand-gated ion channel function ensures that synaptic transmission is both rapid and precise. The receptor's ability to open and close within milliseconds allows for the quick relay of signals, which is critical for processes requiring fast communication, such as reflexes and voluntary muscle movements.

Termination of Signal

The termination of the signal occurs when acetylcholine is broken down by acetylcholinesterase, an enzyme present in the synaptic cleft. This breakdown ensures that the receptor is no longer bound to acetylcholine and that the ion channels close, stopping further depolarization of the postsynaptic membrane.

Functions of nAChRs in the Central Nervous System (CNS)

In the central nervous system, nAChRs play an essential role in a wide variety of cognitive and sensory processes. They are not only involved in synaptic transmission but also influence neural plasticity, attention, learning, and memory.

Cognitive Functions

nAChRs are crucial in modulating cognitive processes such as learning and memory. These receptors are abundant in areas of the brain associated with cognition, particularly the hippocampus and the cerebral cortex. Activation of nAChRs enhances synaptic plasticity, which is essential for the encoding and retrieval of memories. For example, nAChR activation can increase the release of other neurotransmitters like glutamate, which in turn strengthens synaptic connections and facilitates long-term potentiation (LTP), a mechanism underlying memory formation.

Modulation of Attention

The cholinergic system, through nAChRs, also plays a key role in regulating attention. The α4β2 and α7 nAChR subtypes are particularly involved in this process. When activated, these receptors enhance the release of dopamine and other neurotransmitters, which helps maintain alertness and focus. Dysfunction in these receptors has been linked to attention-deficit disorders, and research is ongoing to understand their potential in treating such conditions.

Sensory Processing

nAChRs are involved in sensory processing, especially in areas such as the auditory cortex and the somatosensory cortex. These receptors enhance sensory processing by regulating the release of neurotransmitters, thus fine-tuning the brain's ability to process and respond to external stimuli.

Neuropathological Implications

In neurodegenerative diseases such as Alzheimer's and Parkinson's, cholinergic dysfunction, particularly the dysfunction of nAChRs, is a significant contributing factor. For instance, in Alzheimer's disease, the loss of cholinergic neurons in the basal forebrain reduces the number of nAChRs available for synaptic communication, leading to cognitive deficits.

Functions of nAChRs in the Peripheral Nervous System (PNS)

In the peripheral nervous system, nAChRs are primarily involved in neuromuscular transmission and autonomic regulation. These functions are crucial for everyday movement and the regulation of involuntary bodily functions.

Neuromuscular Junction

One of the most well-known functions of nAChRs is their role in mediating communication between motor neurons and muscle fibers at the neuromuscular junction. The α1β1δγ/ε subunit combination of the receptor is found here, where it facilitates the release of acetylcholine. The binding of acetylcholine to nAChRs on the muscle cell membrane leads to the opening of ion channels and the influx of sodium ions, resulting in muscle depolarization and ultimately muscle contraction.

Autonomic Nervous System Regulation

nAChRs are also essential for synaptic transmission in the autonomic ganglia, where they mediate the communication between pre- and post-ganglionic neurons. These receptors are found in both the sympathetic and parasympathetic nervous systems, where they help regulate functions such as heart rate, digestion, and respiratory rate. Activation of nAChRs in these systems contributes to the appropriate response to various physiological demands, such as increasing heart rate during exercise or promoting digestion during rest.

Muscle Disorders

nAChRs play a key role in the proper functioning of skeletal muscles. Disorders such as myasthenia gravis, an autoimmune condition that targets nAChRs at the neuromuscular junction, result in muscle weakness and fatigue. Understanding the physiological role of nAChRs in this context is critical for the development of therapeutic strategies aimed at restoring normal muscle function.

The Connection Between nAChRs and Motor Function

Motor function is largely dependent on the action of nAChRs at the neuromuscular junction and within the central nervous system. In particular, the coordination of muscle contractions requires precise signaling through these receptors.

Skeletal Muscle Contraction

At the neuromuscular junction, the binding of acetylcholine to nAChRs leads to the opening of ion channels and the subsequent depolarization of the muscle membrane. This depolarization triggers the release of calcium ions from the sarcoplasmic reticulum, which interacts with muscle fibers to induce contraction. Proper functioning of nAChRs is therefore essential for voluntary movement and muscle strength.

Motor Control in the Brain

In the central nervous system, nAChRs contribute to the regulation of motor control. The basal ganglia, which are involved in voluntary motor activity, rely on nAChRs for the modulation of neurotransmitter release, particularly dopamine. Dysfunction of nAChRs in these areas can lead to movement disorders, including those seen in Parkinson's disease, where the loss of dopaminergic input affects motor coordination.

Interaction with Other Neurotransmitter Systems

nAChRs do not operate in isolation; rather, they interact with a variety of other neurotransmitter systems, including dopaminergic, serotonergic, and glutamatergic systems. These interactions are crucial for regulating mood, cognition, and movement.

Dopaminergic System

nAChRs in the brain are involved in modulating the release of dopamine, a neurotransmitter critical for reward, motivation, and motor function. Activation of nAChRs can enhance dopamine release, which is why these receptors are implicated in addiction and mood disorders. Dysregulation of dopamine release due to impaired nAChR function can lead to conditions such as depression and schizophrenia.

Glutamatergic System

nAChRs also interact with the glutamatergic system, which is involved in excitatory synaptic transmission. Activation of certain nAChR subtypes increases the release of glutamate, further enhancing synaptic transmission and plasticity. This interaction is important for learning, memory, and higher cognitive functions.

Serotonergic System

nAChRs also influence serotonin levels, impacting mood regulation and sleep. The serotonergic system plays a significant role in maintaining emotional balance, and nAChRs contribute to this process by modulating serotonin release in various regions of the brain.

Conclusion

Nicotinic acetylcholine receptors (nAChRs) are central to the proper functioning of both the central and peripheral nervous systems. They mediate synaptic transmission, facilitate motor function, and contribute to the regulation of cognitive processes, such as learning and memory. Additionally, their interactions with other neurotransmitter systems underscore their importance in maintaining overall physiological and psychological health. Understanding the physiological roles of nAChRs offers insight into their contribution to both normal

Chapter 5: Pharmacology of Nicotinic Acetylcholine Receptors (nAChRs)

The pharmacology of nicotinic acetylcholine receptors (nAChRs) is a critical aspect of understanding how these receptors function both in normal physiological processes and in the context of diseases and therapeutic interventions. The interaction of nAChRs with various pharmacological agents—whether agonists, antagonists, or modulators—can drastically alter receptor function, influencing a wide range of biological systems. This chapter provides an overview of the pharmacology of nAChRs, focusing on the agonists and antagonists that affect these receptors, the synthetic and natural compounds involved, their therapeutic and toxicological implications, and the role of nAChRs in addiction and dependence.

Overview of nAChR Agonists and Antagonists

The pharmacological properties of nAChRs are defined by their response to agonists and antagonists. Agonists are compounds that bind to the receptor and activate it, mimicking the action of acetylcholine, while antagonists bind to the receptor but do not activate it, blocking the receptor's activity.

Agonists

- **Acetylcholine**: As the endogenous ligand for nAChRs, acetylcholine is the most significant agonist. It binds to the receptor, causing a conformational change that opens the ion channel and initiates synaptic transmission.
- **Nicotine**: Nicotine is one of the most studied exogenous nAChR agonists. It binds to nAChRs with high affinity, particularly the α4β2 and α7 subtypes, and mimics acetylcholine's effects. Nicotine's prolonged activation of nAChRs plays a significant role in addiction and dependence, as it enhances dopamine release in the brain's reward pathways.
- **Other Agonists**: Other synthetic and natural compounds can also act as nAChR agonists. For example, compounds such as varenicline (used in smoking cessation) selectively activate certain nAChR subtypes without causing the full range of nicotine's effects, thus aiding in the reduction of nicotine cravings.

Antagonists

- **Curare**: A well-known antagonist of nAChRs, curare is derived from certain plant species and has been used historically as a paralytic agent. It binds to nAChRs at the neuromuscular junction, preventing acetylcholine from activating the receptor, and leading to muscle paralysis.
- **α-Bungarotoxin**: A potent and selective antagonist, α-bungarotoxin binds irreversibly to nAChRs, particularly those in the central nervous system, and is often used in research to study receptor function.
- **Other Antagonists**: Many synthetic drugs, such as mecamylamine, act as antagonists to nAChRs, blocking acetylcholine's ability to activate the receptor. These drugs are studied for their potential therapeutic use in treating conditions such as hypertension, psychiatric disorders, and addiction.

Synthetic and Natural Compounds Affecting nAChRs

In addition to nicotine, there are several synthetic and natural compounds that can interact with nAChRs, either by directly binding to the receptors or by influencing their expression or function in other ways.

Natural Compounds

- **Nicotine**: The most prominent natural compound that affects nAChRs, nicotine not only stimulates nAChRs but also causes long-term changes in their expression, contributing to addiction. Its interaction with nAChRs in the brain's reward system is central to the development of dependence.
- **Caffeine**: While caffeine primarily acts as an adenosine receptor antagonist, it has been shown to modulate nAChR activity, particularly in the context of learning and memory, by increasing dopamine release and enhancing synaptic transmission in certain brain regions.
- **Alkaloids from Tobacco and Other Plants**: Other alkaloids, such as those found in tobacco, also interact with nAChRs, leading to similar effects as nicotine, including receptor desensitization and the modulation of neurotransmitter release.

Synthetic Compounds

- **Varenicline (Chantix)**: Varenicline is a partial agonist of nAChRs that selectively binds to the α4β2 subtype, providing enough activation to reduce nicotine cravings without causing the full addictive effects of nicotine. It is used as a smoking cessation aid, offering a promising therapeutic option with fewer side effects than nicotine replacement therapies.
- **Bupropion (Zyban)**: Bupropion, an atypical antidepressant, also influences nAChRs. It is believed to act as both a dopamine reuptake inhibitor and an nAChR antagonist, helping to reduce cravings and withdrawal symptoms in smokers trying to quit.

Therapeutic and Toxicological Implications

The pharmacological activity of nAChRs has profound therapeutic and toxicological implications. The ability to modulate nAChR activity has applications in treating a wide range of diseases, including neurological disorders, addiction, and mental health conditions. However, the same properties that make nAChRs a valuable target for therapy also make them a potential source of toxicity.

Therapeutic Implications

- **Neurological Diseases**: Drugs that target nAChRs are being investigated for their therapeutic potential in treating neurodegenerative diseases such as Alzheimer's disease and Parkinson's disease. For example, certain nAChR agonists have shown promise in enhancing cognitive function by promoting synaptic plasticity and enhancing neurotransmitter release, particularly in the cholinergic pathways that are often compromised in these diseases.
- **Smoking Cessation**: Varenicline and other nicotinic modulators provide an effective pharmacological approach to smoking cessation. These agents reduce cravings and withdrawal symptoms associated with nicotine dependence while minimizing the reinforcement of addictive behavior.
- **Mental Health Disorders**: nAChR antagonists are being studied for their potential in treating conditions like schizophrenia and depression. Modulating nAChRs in the central nervous system could help balance neurotransmitter systems that are out of equilibrium in these disorders, potentially improving cognitive function and mood regulation.

Toxicological Implications

- **Nicotine Poisoning**: Excessive nicotine exposure can lead to nicotine toxicity, causing symptoms such as nausea, vomiting, dizziness, confusion, and, in severe cases, seizures and respiratory failure. This toxicity occurs due to the overstimulation of nAChRs, resulting in excessive ion flux and disturbances in neuronal communication.
- **Curare and Other nAChR Antagonists**: Drugs that block nAChRs, such as curare, can be toxic by inhibiting neuromuscular transmission and causing paralysis. While these agents have clinical applications as muscle relaxants during surgery, they must be carefully dosed to prevent lethal outcomes.

The Role of nAChRs in Addiction and Dependence

The role of nAChRs in addiction is most prominently illustrated by the addictive properties of nicotine, the primary psychoactive compound in tobacco. Nicotine's ability to activate nAChRs, particularly in the brain's reward pathways, leads to the release of dopamine, reinforcing the behavior of tobacco use.

Nicotine Addiction

Nicotine's interaction with nAChRs in the mesolimbic dopaminergic system, which is involved in reward and motivation, is central to the development of nicotine addiction. Chronic exposure to nicotine causes changes in nAChR expression, leading to tolerance and dependence. As a result, individuals who attempt to quit smoking may experience withdrawal symptoms, including irritability, anxiety, and difficulty concentrating.

Other Addictions

Research has suggested that nAChRs may also play a role in the addiction to other substances, such as cocaine, alcohol, and opioids. nAChR modulation can influence the release of dopamine and other neurotransmitters involved in addiction pathways, suggesting that targeting these receptors could be a therapeutic strategy for managing a range of substance use disorders.

Therapeutic Strategies for Addiction

Agents such as varenicline, which partially activate nAChRs, provide a potential treatment for nicotine addiction by reducing cravings and withdrawal symptoms while helping to prevent the reinforcing effects of smoking. Other potential therapies targeting nAChRs are being explored for alcohol and cocaine addiction, with the goal of reducing the pleasurable effects of these substances without causing full receptor activation.

Conclusion

The pharmacology of nAChRs is integral to understanding their function in both health and disease. The diverse range of nAChR agonists and antagonists, from natural compounds like nicotine to synthetic drugs like varenicline, underscores the potential for modulating these receptors for therapeutic benefit. However, the same properties that make nAChRs an attractive drug target also present significant challenges in terms of toxicity and addiction. By exploring the pharmacological mechanisms of nAChRs, we gain valuable insight into their therapeutic and toxicological implications, paving the way for the development of novel treatments for addiction, neurological disorders, and other conditions.

Summary

In this chapter, we have explored the pharmacology of nicotinic acetylcholine receptors (nAChRs), including the effects of both synthetic and natural compounds on receptor function. The therapeutic and toxicological implications of nAChR modulation are profound, offering potential treatments for addiction, neurological diseases, and mental health conditions. However, the addictive nature of nicotine and the risks associated with nAChR antagonism also highlight the need for careful management in both clinical and research settings. The role of nAChRs in addiction, particularly nicotine dependence, further emphasizes the importance of these receptors in human health.

Chapter 6: nAChRs in the Central Nervous System

The central nervous system (CNS) is a complex network of neurons that rely on precise and rapid signaling to maintain bodily functions, cognition, and behavior. Nicotinic acetylcholine receptors (nAChRs) play a crucial role in the CNS, influencing a wide variety of processes, from basic synaptic transmission to higher-order cognitive functions. This chapter will explore the role of nAChRs in the CNS, particularly their involvement in cognitive processes, neurodegenerative diseases, mood regulation, and neuropsychological disorders.

Function in Cognitive Processes: Learning and Memory

nAChRs are widely distributed throughout the CNS and are especially abundant in regions responsible for cognitive functions, such as the hippocampus, cortex, and basal ganglia. These receptors are involved in modulating neural activity and are essential for synaptic plasticity—the ability of synapses to strengthen or weaken in response to activity. Synaptic plasticity is a fundamental process underlying learning and memory formation.

Synaptic Plasticity and Long-Term Potentiation (LTP)

One of the most well-established roles of nAChRs in the CNS is their involvement in synaptic plasticity, particularly in the hippocampus and cortex, areas known for their roles in learning and memory. Activation of nAChRs promotes the release of neurotransmitters such as glutamate, which in turn activates glutamatergic receptors that facilitate long-term potentiation (LTP)—the process that strengthens synapses in response to repeated stimuli. LTP is thought to be a cellular mechanism of memory and learning, and dysfunction in this process has been associated with cognitive impairments.

Enhancement of Cognitive Function

nAChRs contribute to cognitive enhancement by regulating attention, focus, and information processing. The activation of α4β2 and α7 subtypes of nAChRs in the prefrontal cortex has been shown to enhance working memory and attention. These effects are particularly important for tasks requiring sustained mental effort, such as problem-solving and decision-making.

Role in Attention

nAChRs also play a key role in attention and vigilance. Activation of nAChRs in the cortex and brainstem increases the release of acetylcholine and other neurotransmitters such as dopamine, which enhance focus and sensory processing. This interaction is essential for filtering out irrelevant stimuli and concentrating on relevant tasks. Impaired nAChR function in these regions can lead to deficits in attention, which are observed in disorders such as ADHD.

Involvement in Neurodegenerative Diseases: Alzheimer's, Parkinson's

The dysfunction of nAChRs in the CNS has been implicated in several neurodegenerative diseases, particularly Alzheimer's disease and Parkinson's disease. These diseases are characterized by the progressive loss of neurons and the disruption of neurotransmitter systems, including cholinergic signaling.

Alzheimer's Disease

Alzheimer's disease (AD) is characterized by a decline in cognitive function, particularly in memory and executive function. One of the hallmark features of AD is the degeneration of cholinergic neurons in the basal forebrain, which leads to a reduction in acetylcholine and a subsequent loss of nAChRs in the cortex and hippocampus. This cholinergic dysfunction impairs synaptic plasticity, which is critical for learning and memory. Researchers are investigating nAChR agonists and other cholinergic modulators as potential treatments for AD, with the aim of improving cognitive function by enhancing synaptic transmission.

Parkinson's Disease

Parkinson's disease (PD) is primarily characterized by the degeneration of dopaminergic neurons in the substantia nigra, leading to motor deficits such as tremors, rigidity, and bradykinesia. However, nAChRs also play a critical role in PD. Dopamine depletion in the basal ganglia alters the balance between excitatory and inhibitory neurotransmission, and nAChRs in the striatum and cortex can modulate dopaminergic activity. Stimulation of nAChRs may help restore some of this balance, and nAChR agonists are being explored as potential adjuncts to traditional PD therapies.

Therapeutic Potential in Neurodegeneration

The role of nAChRs in neurodegenerative diseases underscores their therapeutic potential. Agonists that target specific nAChR subtypes, such as α7 and α4β2, are being studied for their ability to restore cholinergic function and improve cognition in diseases like Alzheimer's and Parkinson's. Early research shows promise for these compounds in improving memory and attention in animal models of neurodegeneration.

Effects on Mood, Anxiety, and Depression

nAChRs in the brain also regulate emotional states, including mood, anxiety, and depression. These receptors are involved in the release of several neurotransmitters, including serotonin, dopamine, and gamma-aminobutyric acid (GABA), which play central roles in regulating mood and emotional responses.

Mood Regulation and Dopamine Release

Activation of nAChRs in regions such as the prefrontal cortex and ventral tegmental area (VTA) can increase dopamine release, which is associated with the regulation of mood and motivation. Dysfunctional dopaminergic signaling is implicated in various mood disorders, including depression and bipolar disorder. nAChRs' ability to modulate dopamine release suggests that these receptors may be a target for antidepressant therapies.

Anxiety

Research has shown that nAChRs, particularly the α7 subtype, play a role in regulating anxiety. The activation of these receptors in the amygdala and other regions involved in fear processing can modulate the stress response. Dysfunction in nAChR signaling in these areas may contribute to anxiety disorders. Thus, nAChR agonists or modulators have potential as anxiolytic agents.

Depression

nAChRs are also implicated in depression, particularly through their interaction with the serotonergic system. Studies have shown that nAChR activation can increase the release of serotonin, a neurotransmitter that regulates mood, sleep, and appetite. Disruption in this system is common in individuals with depression. Therefore, targeting nAChRs, either through direct agonists or by modulating their activity, could provide a novel avenue for treating depression.

Research on nAChRs in Neuropsychological Disorders

The role of nAChRs in various neuropsychological disorders, including schizophrenia, ADHD, and mood disorders, has been an active area of research. The dysfunction of these receptors may contribute to the symptoms of these conditions, and understanding their involvement can lead to new treatment strategies.

Schizophrenia

nAChRs are involved in modulating dopaminergic and glutamatergic transmission, both of which are implicated in the pathophysiology of schizophrenia. Studies have shown that nAChR abnormalities, particularly in the hippocampus and cortex, may contribute to cognitive deficits and other symptoms of schizophrenia. Investigating nAChR function and developing drugs that target these receptors could offer new therapeutic strategies for managing the cognitive and negative symptoms of schizophrenia.

Attention-Deficit/Hyperactivity Disorder (ADHD)

nAChRs are also involved in attention and focus, and their dysfunction may contribute to the symptoms of ADHD. Research has indicated that nAChR agonists could have potential as therapeutic agents in improving attention and reducing hyperactivity in ADHD. The ability of nAChRs to modulate dopamine and norepinephrine release makes them an attractive target for ADHD treatments.

Post-Traumatic Stress Disorder (PTSD)

There is growing interest in the role of nAChRs in post-traumatic stress disorder (PTSD), where dysregulation of the stress response system is a key feature. nAChR modulation could potentially alter the brain's response to stress and reduce the severity of PTSD symptoms. Clinical trials exploring nAChR-targeting agents for PTSD are ongoing.

Conclusion

Nicotinic acetylcholine receptors (nAChRs) play a pivotal role in the central nervous system, influencing a wide array of functions from cognitive performance and mood regulation to the pathophysiology of neurodegenerative and neuropsychological disorders. Their involvement in diseases such as Alzheimer's and Parkinson's highlights their therapeutic potential, while their role in mood disorders, anxiety, and depression presents an exciting area for future research. Understanding the complex interactions of nAChRs in the CNS offers opportunities for the development of novel therapeutic strategies that could improve the quality of life for individuals suffering from these challenging conditions.

Summary

This chapter has provided an overview of the critical role of nAChRs in the central nervous system, focusing on their influence in cognitive functions, neurodegenerative diseases, mood regulation, and neuropsychological disorders. By exploring the effects of nAChRs on learning, memory, anxiety, and neurodegeneration, we have highlighted their potential as therapeutic targets for a variety of conditions. In the next chapter, we will examine the role of nAChRs in the peripheral nervous system and their involvement in muscular function, autonomic regulation, and pain modulation.

Chapter 7: nAChRs in the Peripheral Nervous System

Nicotinic acetylcholine receptors (nAChRs) are fundamental to the functioning of both the central and peripheral nervous systems. While their role in the central nervous system (CNS) has been the focus of much research, nAChRs in the peripheral nervous system (PNS) are equally critical for various physiological processes, including neuromuscular transmission, autonomic regulation, and pain modulation. This chapter will explore the roles of nAChRs in the peripheral nervous system, focusing on their contribution to autonomic nervous system regulation, muscular function, neuromuscular diseases, and pain modulation.

Autonomic Nervous System Regulation

The autonomic nervous system (ANS) controls involuntary bodily functions such as heart rate, blood pressure, digestion, and respiration. It consists of two primary branches: the sympathetic and parasympathetic nervous systems. nAChRs are involved in the synaptic transmission of signals within both branches, playing a vital role in regulating these essential physiological processes.

Parasympathetic Nervous System

- In the parasympathetic branch, nAChRs are involved in transmitting signals from preganglionic neurons to postganglionic neurons at autonomic ganglia. The activation of nAChRs in these ganglia allows for the communication between pre- and post-ganglionic neurons, contributing to various parasympathetic functions such as slowing the heart rate, stimulating digestion, and promoting rest-and-digest activities.
- The α3β4 nAChR subtype is particularly prominent in the parasympathetic nervous system and is involved in regulating functions like gastrointestinal motility and secretions. Dysfunction in nAChR signaling in the parasympathetic system can result in disorders related to heart rate regulation, gastrointestinal motility, and other autonomic functions.

Sympathetic Nervous System

- Similarly, nAChRs in the sympathetic nervous system help mediate the transmission of signals from pre- to post-ganglionic neurons. The sympathetic system is primarily responsible for the fight-or-flight response, which involves the activation of the heart, lungs, and other organs in response to stress.
- The β2 subtype of nAChRs is often found in sympathetic ganglia, where it plays a role in regulating blood pressure and vasoconstriction. Dysregulation of nAChRs in this context can lead to cardiovascular issues, such as hypertension and impaired stress responses.

In summary, nAChRs play an integral role in regulating autonomic functions by enabling the communication between pre- and post-ganglionic neurons. By mediating the transmission of signals that control heart rate, digestion, and stress responses, nAChRs ensure that the autonomic nervous system maintains homeostasis in the body.

Muscular Function and nAChRs in Motor Neurons

One of the most well-known roles of nAChRs is their involvement in neuromuscular transmission, which is crucial for muscle contraction. At the neuromuscular junction (NMJ), nAChRs mediate the communication between motor neurons and muscle fibers, a process essential for voluntary movement.

Neuromuscular Junction

- At the neuromuscular junction, the motor neuron releases acetylcholine (ACh), which binds to nAChRs on the muscle cell membrane (sarcolemma). This binding opens ion channels, allowing sodium (Na+) ions to enter the muscle cell, which leads to depolarization. This depolarization triggers the release of calcium ions (Ca2+) from the sarcoplasmic reticulum, initiating muscle contraction. Once acetylcholine is broken down by acetylcholinesterase, the ion channels close, and muscle contraction ceases.
- The nAChR in the NMJ is typically composed of α1, β1, δ, and γ or ε subunits (the γ subunit being present in fetal tissue and transitioning to the ε subunit in adulthood). The presence of the α1 subunit is crucial for acetylcholine binding and receptor activation, making it a target for drugs that modulate muscle function.

Muscle Contraction and Control

The efficiency of neuromuscular transmission directly impacts muscle strength and coordination. When nAChRs function properly, the muscles can contract efficiently in response to nerve signals. However, impaired nAChR function or reduced receptor availability can lead to weakened muscle responses and coordination deficits.

Disorders of Neuromuscular Function

- Several neuromuscular diseases are associated with dysfunctional nAChRs. One such condition is **myasthenia gravis**, an autoimmune disorder in which antibodies attack nAChRs at the neuromuscular junction, leading to muscle weakness and fatigue. In myasthenia gravis, the reduced number of functioning nAChRs means that muscle fibers cannot be activated effectively, resulting in impaired voluntary movement.
- **Congenital myasthenic syndromes** (CMS) are genetic conditions that also involve defects in nAChR function, leading to muscle weakness from birth or early childhood. These conditions may involve mutations in the genes encoding nAChR subunits, leading to problems with receptor assembly, function, or stability.

In summary, nAChRs at the neuromuscular junction are critical for muscle contraction, and any dysfunction in these receptors can lead to neuromuscular diseases, such as myasthenia gravis or congenital myasthenic syndromes, both of which result in debilitating muscle weakness and loss of motor control.

nAChRs and Pain Modulation

The role of nAChRs in pain modulation is an emerging area of research. nAChRs are expressed in various parts of the peripheral nervous system, including sensory neurons and pain pathways. Their activation has been shown to have both pro-nociceptive (pain-enhancing) and anti-nociceptive (pain-reducing) effects, depending on the context and the specific nAChR subtypes involved.

Peripheral Sensory Nerve Activation

nAChRs are present on sensory neurons that detect noxious stimuli (pain) and transmit pain signals to the brain. Activation of these receptors can enhance the release of neurotransmitters such as substance P and glutamate, which are involved in pain transmission. However, in some cases, activation of certain nAChRs, such as the α7 subtype, has been shown to inhibit the release of these neurotransmitters, reducing the perception of pain.

Anti-Nociceptive Effects

The activation of α7 nAChRs in the peripheral nervous system may lead to anti-nociceptive effects, offering potential for the development of new pain management strategies. By modulating nAChR signaling, it may be possible to enhance the body's natural ability to inhibit pain transmission without the need for traditional analgesics or opioids, thus avoiding issues of dependency and side effects.

Inflammation and Pain

nAChRs are also involved in modulating inflammation, which can influence pain perception. The activation of nAChRs in immune cells, such as macrophages, can reduce the release of pro-inflammatory cytokines, which are often associated with pain and inflammation. This suggests that nAChR-based therapies may not only help to alleviate pain directly but also reduce the underlying inflammation contributing to chronic pain conditions.

Pain in Neuropathic Disorders

Neuropathic pain, resulting from damage to sensory nerves, is often difficult to treat with traditional analgesics. Emerging research suggests that nAChR modulation may offer new avenues for treating neuropathic pain. For instance, nAChR agonists have been shown to reduce pain in animal models of neuropathic pain, highlighting their potential for use in clinical settings.

In conclusion, nAChRs have a significant role in pain modulation, with their activation influencing both the perception of pain and the underlying inflammatory processes that contribute to chronic pain. As research progresses, targeting nAChRs in the peripheral nervous system may offer new therapeutic strategies for pain management.

Conclusion

Nicotinic acetylcholine receptors (nAChRs) in the peripheral nervous system are central to a variety of physiological processes, including the regulation of autonomic functions, muscular contraction, and pain perception. Their involvement in neuromuscular transmission makes them essential for voluntary movement, while their role in autonomic nervous system regulation ensures homeostasis across multiple bodily systems. The dysfunction of these receptors can lead to debilitating conditions, such as myasthenia gravis, and their modulation offers exciting potential for the treatment of chronic pain and other neuromuscular disorders. Understanding the physiology of nAChRs in the peripheral nervous system will be critical for the development of targeted therapies for these conditions.

Summary

This chapter has explored the role of nAChRs in the peripheral nervous system, focusing on their contribution to autonomic nervous system regulation, muscular function, neuromuscular diseases, and pain modulation. By mediating crucial functions in the body's involuntary systems and facilitating muscle contraction, nAChRs are essential for maintaining bodily homeostasis and motor control. Their dysfunction can lead to neuromuscular diseases and impaired pain regulation, making them a key target for therapeutic interventions in these areas. In the next chapter, we will examine the role of nAChRs in addiction, particularly nicotine addiction, and the potential for therapeutic strategies targeting nAChRs to manage substance use disorders.

Chapter 8: nAChRs and Addiction

Addiction is a complex and multifactorial condition characterized by compulsive drug-seeking behavior, loss of control over substance use, and continued use despite harmful consequences. Nicotinic acetylcholine receptors (nAChRs) play a central role in addiction, particularly nicotine addiction, but also in the addiction to other substances such as cocaine, alcohol, and opioids. This chapter will explore the mechanisms through which nAChRs contribute to addiction, focusing on nicotine addiction, the role of nAChRs in other forms of substance abuse, therapeutic strategies for managing addiction, and ongoing research into treatments that target nAChRs.

Nicotine Addiction: Mechanisms and Pathways

Nicotine, the primary addictive substance in tobacco, is a powerful agonist of nAChRs. Its addictive properties are largely due to its ability to rapidly stimulate nAChRs, which leads to the release of dopamine, a neurotransmitter associated with pleasure and reward, in the brain's mesolimbic system.

Dopaminergic Pathways

- When nicotine binds to nAChRs, particularly the α4β2 and α7 subtypes, it stimulates the release of dopamine in the mesolimbic dopamine pathway, which includes the nucleus accumbens and ventral tegmental area (VTA). This dopamine release is part of the brain's reward system and plays a key role in reinforcing the behavior of smoking.
- The release of dopamine creates feelings of pleasure and reinforcement, leading individuals to repeat the behavior. Over time, repeated nicotine exposure leads to neuroplastic changes in the brain, such as the upregulation of nAChRs and increased dopamine receptor expression. These changes contribute to tolerance, dependence, and the development of addiction.

Neuroplasticity and Sensitization

Chronic nicotine use leads to long-term alterations in nAChR expression, particularly in the brain's reward circuits. This process of neuroplasticity, which involves changes in receptor density and synaptic activity, plays a major role in addiction. Over time, individuals require more nicotine to achieve the same effect, which is a hallmark of tolerance. Additionally, these changes in nAChRs lead to increased sensitivity to nicotine and other drugs, contributing to the compulsive nature of addiction.

Withdrawal and Craving

When a person attempts to quit smoking, nicotine withdrawal occurs, and it is often accompanied by symptoms such as irritability, anxiety, depression, difficulty concentrating, and cravings for nicotine. These withdrawal symptoms are primarily driven by the decreased availability of dopamine and altered nAChR function in the brain. The craving for nicotine is reinforced by the brain's reward pathways, making relapse a common outcome for many individuals trying to quit.

The Role of nAChRs in Drug Abuse (Cocaine, Alcohol, etc.)

While nicotine is the most studied substance in the context of nAChR involvement in addiction, other drugs of abuse, including cocaine, alcohol, and opioids, also interact with nAChRs, contributing to the addictive properties of these substances.

Cocaine

Cocaine increases dopamine release by inhibiting the reuptake of dopamine in the synapse. Research has shown that nAChRs, particularly the α4β2 subtype, modulate dopamine release in response to cocaine. Nicotine exposure can increase the rewarding effects of cocaine, potentially making individuals more susceptible to cocaine addiction. Additionally, nAChRs play a role in the enhancement of dopamine signaling in response to cocaine, further reinforcing drug-seeking behavior.

Alcohol

Alcohol is another substance that interacts with nAChRs, particularly in the brain's reward pathways. Alcohol enhances the release of dopamine and serotonin, both of which are involved in mood regulation and reward. nAChRs, particularly those in the VTA and nucleus accumbens, modulate the effects of alcohol on dopamine release, thus contributing to alcohol's reinforcing effects. Studies have shown that blocking nAChRs can reduce alcohol consumption, suggesting that nAChRs play a significant role in alcohol addiction.

Opioids

Opioids, such as heroin and morphine, exert their addictive effects primarily through the opioid receptors. However, there is evidence that nAChRs also play a role in opioid addiction. nAChRs, particularly those in the mesolimbic dopamine pathway, modulate the release of dopamine in response to opioid use. This suggests that nAChRs may be involved in the reinforcing effects of opioids and may contribute to the development of opioid dependence.

Cannabis

While cannabis primarily affects the endocannabinoid system, there is evidence that nAChRs also play a role in its effects. Studies have shown that nicotine and cannabis use can influence each other's effects on dopamine release, highlighting the complex interactions between nAChRs and other neurotransmitter systems involved in addiction.

Therapeutic Strategies for Addiction Management

Given the central role of nAChRs in addiction, targeting these receptors offers a promising therapeutic approach for managing nicotine and other substance use disorders. Current therapeutic strategies focus on modulating nAChR activity to reduce cravings, withdrawal symptoms, and the reinforcing effects of drugs.

Nicotine Replacement Therapy (NRT)

NRT, which includes products such as nicotine gums, patches, lozenges, and inhalers, aims to provide a controlled, lower dose of nicotine to reduce withdrawal symptoms and cravings while helping individuals quit smoking. NRT works by stimulating nAChRs to some extent, thus easing the transition from smoking to quitting. However, it does not fully activate nAChRs in the same way smoking does, reducing the risk of reinforcing addiction.

Varenicline (Chantix)

Varenicline is a partial agonist of the α4β2 nAChR subtype and is used as a smoking cessation aid. It stimulates nAChRs to a lesser extent than nicotine, which reduces cravings and withdrawal symptoms. Varenicline also blocks nicotine from binding to nAChRs, which helps prevent the reinforcing effects of smoking. Clinical trials have shown that varenicline is more effective than placebo and comparable to other forms of NRT in helping people quit smoking.

Bupropion (Zyban)

Bupropion is another drug used to help individuals quit smoking. While it primarily works as a dopamine reuptake inhibitor and norepinephrine reuptake inhibitor, it also affects nAChRs by modulating dopamine release in the brain. This helps to reduce nicotine cravings and withdrawal symptoms. Bupropion has also shown efficacy in treating other forms of addiction, including cocaine and alcohol dependence.

Research on Other nAChR Modulators

Ongoing research is focused on developing new drugs that more selectively target nAChRs to treat various forms of addiction. These include drugs that act as partial agonists or antagonists to modulate nAChR activity without inducing full activation. For example, novel compounds that target the α7 or α4β2 nAChR subtypes are being explored for their ability to reduce the rewarding effects of nicotine, cocaine, alcohol, and other addictive substances.

Behavioral Therapies

While pharmacological treatments are important, behavioral therapies, such as cognitive-behavioral therapy (CBT) and contingency management, are also effective in treating addiction. These therapies can help individuals develop coping strategies to manage cravings and triggers associated with addiction. Combining pharmacological treatments that target nAChRs with behavioral therapies provides a more comprehensive approach to addiction management.

Research on Potential Treatments Targeting nAChRs

The study of nAChRs in addiction has opened up exciting possibilities for new treatments. Researchers are investigating selective nAChR modulators that can target specific receptor subtypes involved in different forms of addiction. Advances in molecular pharmacology, as well as the use of animal models and human clinical trials, will continue to drive the development of novel therapeutic strategies. Additionally, the rise of precision medicine and genetic screening may allow for more personalized treatments that are tailored to an individual's specific nAChR genetic profile.

Selective nAChR Agonists and Antagonists

One area of active research involves developing drugs that can selectively activate or block specific nAChR subtypes involved in addiction. For example, drugs targeting the α7 nAChR subtype may help reduce the reinforcing effects of nicotine and other drugs without causing full receptor activation.

Gene Therapy and CRISPR Technology

The use of gene therapy and CRISPR technology to modify nAChR expression is still in the early stages but holds promise for the future. By directly targeting the genes responsible for nAChR function, researchers may be able to alter the brain's response to addictive substances and help individuals overcome addiction.

Conclusion

Nicotinic acetylcholine receptors (nAChRs) play a central role in the development and reinforcement of addiction. Their ability to modulate dopamine release in response to substances such as nicotine, cocaine, and alcohol makes them key players in addictive behavior. Understanding the mechanisms by which nAChRs contribute to addiction has led to the development of several therapeutic strategies aimed at modulating these receptors, including nicotine replacement therapies, varenicline, and bupropion. As research continues to explore the potential for selective nAChR modulators and genetic therapies, the future of addiction treatment looks promising. By targeting nAChRs in combination with behavioral therapies, it may be possible to develop more effective, personalized treatments for addiction and improve the lives of individuals struggling with substance use disorders.

Chapter 9: The Role of nAChRs in Neurological Disorders

Nicotinic acetylcholine receptors (nAChRs) are integral to the normal functioning of the nervous system, and their dysregulation is implicated in various neurological disorders. From neurodegenerative diseases to psychiatric conditions, the dysfunction of nAChRs can contribute to a broad spectrum of cognitive, motor, and psychological symptoms. This chapter explores the role of nAChRs in neurological disorders such as Alzheimer's disease, Parkinson's disease, Huntington's disease, and schizophrenia, highlighting their potential as therapeutic targets and the challenges in developing treatments based on nAChR modulation.

Alzheimer's Disease: Cholinergic Dysfunction and nAChRs

Alzheimer's disease (AD) is characterized by progressive cognitive decline, particularly in memory, and is the most common form of dementia in older adults. The disease is marked by the accumulation of amyloid plaques and tau tangles in the brain, leading to the degeneration of neurons, especially those in the basal forebrain, a region rich in cholinergic neurons.

Cholinergic Dysfunction in AD

One of the hallmark features of AD is the loss of cholinergic neurons in the basal forebrain, which results in a decrease in acetylcholine (ACh) levels in the cortex and hippocampus. This depletion of acetylcholine disrupts the normal functioning of nAChRs, particularly those in the hippocampus, which are critical for learning and memory. As a result, individuals with AD experience significant cognitive impairment, especially in memory formation and retrieval.

nAChRs as Therapeutic Targets

nAChRs, particularly the α4β2 and α7 subtypes, are attractive targets for therapeutic intervention in AD. Studies suggest that these receptors play a key role in cognitive functions like synaptic plasticity, attention, and memory. Agonists or modulators that enhance nAChR activity could restore cholinergic signaling in the brain and improve cognitive function in AD patients. While some nAChR-targeting compounds have shown promise in preclinical studies, clinical trials have had mixed results, highlighting the complexity of developing effective treatments for AD.

Parkinson's Disease: Dopamine and nAChR Interactions

Parkinson's disease (PD) is a neurodegenerative disorder characterized by the progressive loss of dopaminergic neurons in the substantia nigra, leading to motor symptoms such as tremors, rigidity, and bradykinesia (slowness of movement). While the primary pathophysiology of PD involves dopaminergic dysfunction, nAChRs also play an important role in modulating the motor symptoms associated with the disease.

nAChRs in the Basal Ganglia

The basal ganglia, a group of structures involved in motor control, are rich in nAChRs, particularly α4β2 and α7 subtypes. These receptors influence dopamine release and the activity of dopaminergic neurons in response to acetylcholine. In PD, where dopaminergic neurons are lost, nAChRs in the basal ganglia can help modulate the remaining dopaminergic activity and partially compensate for dopamine depletion.

nAChRs in Motor Control

Activation of nAChRs in the basal ganglia can influence motor activity by enhancing synaptic plasticity and modulating neural circuit function. In animal models of PD, nAChR agonists have shown promise in improving motor symptoms, suggesting that nAChRs might be targeted to improve motor control in PD patients.

Therapeutic Potential

There is growing interest in developing nAChR-targeted therapies for PD, particularly those that can modulate cholinergic signaling in the basal ganglia. Both nAChR agonists and allosteric modulators have been investigated for their potential to restore dopaminergic activity and alleviate PD symptoms. However, challenges remain in balancing the effects on nAChRs to avoid unwanted side effects such as dyskinesias (abnormal involuntary movements) that can occur with dopaminergic therapies.

Huntington's Disease and the Cholinergic System

Huntington's disease (HD) is a genetic neurodegenerative disorder caused by a mutation in the huntingtin gene, leading to progressive motor dysfunction, cognitive decline, and psychiatric symptoms. The disease primarily affects the striatum, a component of the basal ganglia, and the cortex.

Cholinergic Dysfunction in HD

In HD, there is a loss of both GABAergic (inhibitory) and cholinergic neurons in the striatum, leading to dysregulation of motor control and cognitive dysfunction. While the precise role of nAChRs in HD is still being studied, research suggests that alterations in cholinergic signaling may contribute to both the motor and cognitive deficits observed in patients with HD. The loss of cholinergic input may exacerbate the neurodegeneration associated with the disease.

nAChRs in HD

nAChRs, particularly α4β2 and α7 subtypes, are involved in modulating the release of neurotransmitters such as dopamine and glutamate in the striatum. Dysregulation of these receptors may contribute to the movement disorders and cognitive decline seen in HD. Restoration of nAChR function may hold therapeutic potential for ameliorating some of the symptoms of the disease.

Therapeutic Strategies

As with PD, the use of nAChR agonists or modulators has been explored as a potential treatment for HD. Preclinical models have shown that stimulating nAChRs can enhance striatal function and potentially alleviate some of the cognitive and motor deficits associated with HD. However, further research is needed to determine the safety and efficacy of nAChR-targeted therapies for HD patients.

Schizophrenia and the Role of nAChRs in Psychosis

Schizophrenia is a severe mental disorder characterized by symptoms such as delusions, hallucinations, and cognitive impairment. It is thought to involve disruptions in dopamine, glutamate, and serotonin signaling, but cholinergic dysfunction, particularly involving nAChRs, also plays a role in the pathophysiology of the disease.

nAChRs in the CNS

nAChRs are widely distributed in the brain, with high concentrations in areas such as the prefrontal cortex, hippocampus, and basal ganglia, regions that are critically involved in cognition, attention, and emotion regulation. In schizophrenia, there is evidence of abnormal nAChR functioning, particularly in the α7 subtype. Dysfunction in these receptors may contribute to the cognitive and attention deficits that are common in schizophrenia.

Cognitive Impairment in Schizophrenia

One of the most debilitating aspects of schizophrenia is cognitive impairment, including deficits in working memory, attention, and executive function. nAChRs play a crucial role in cognitive processes, and their dysfunction may contribute to these impairments. Additionally, individuals with schizophrenia often smoke at higher rates than the general population, which may be an attempt to self-medicate by activating nAChRs and alleviating some of the cognitive symptoms.

Therapeutic Potential

nAChR agonists, particularly those that selectively target the α7 subtype, are being investigated for their potential to improve cognitive function in schizophrenia. Early clinical trials have shown some promise in improving cognitive symptoms with the use of nAChR modulators. These treatments may offer a novel approach to treating the cognitive deficits associated with schizophrenia, which are typically resistant to current antipsychotic medications.

Conclusion

Nicotinic acetylcholine receptors (nAChRs) play a significant role in the pathophysiology of a variety of neurological disorders, including Alzheimer's disease, Parkinson's disease, Huntington's disease, and schizophrenia. The dysfunction of these receptors contributes to the cognitive, motor, and psychiatric symptoms associated with these conditions. As a result, nAChRs present an attractive target for therapeutic intervention. While nAChR-based therapies have shown promise in preclinical and clinical studies, challenges remain in translating these findings into effective treatments. Continued research into the role of nAChRs in neurological disorders and the development of selective nAChR modulators will be critical in advancing treatment options for patients suffering from these debilitating conditions.

Summary

This chapter has explored the critical role of nAChRs in various neurological disorders, including Alzheimer's disease, Parkinson's disease, Huntington's disease, and schizophrenia. In each of these conditions, nAChR dysfunction contributes to the symptoms observed, and nAChRs offer promising targets for therapeutic development. Future research will be pivotal in understanding the precise mechanisms by which nAChRs influence these diseases and in developing effective treatments that target these receptors to improve patient outcomes.

Chapter 10: Advances in nAChR Research

Nicotinic acetylcholine receptors (nAChRs) have been studied for several decades, with a wealth of information accumulated regarding their structure, function, and involvement in various physiological and pathological processes. However, the field of nAChR research continues to evolve, with ongoing advances that are broadening our understanding of their roles in health and disease. This chapter will explore the latest techniques in nAChR research, including cutting-edge technologies like CRISPR and optogenetics, the critical role of animal models in advancing our understanding of nAChRs, and the potential future directions of nAChR research and its therapeutic implications.

Cutting-Edge Techniques in nAChR Research: CRISPR, Optogenetics, and Beyond

CRISPR Technology

- The CRISPR-Cas9 gene-editing technology has revolutionized molecular biology by allowing for precise and targeted modifications of the genome. In the context of nAChR research, CRISPR is being used to investigate the roles of specific nAChR subtypes and to explore how genetic mutations in nAChR genes contribute to disease.
- Researchers can now use CRISPR to knock out or modify specific nAChR subunits in animal models, providing insights into how changes in receptor expression or function affect neuronal signaling and behavior. This has been particularly useful in studying neurodegenerative diseases, such as Alzheimer's disease and Parkinson's disease, where nAChR dysfunction plays a key role.

Optogenetics

- Optogenetics is another cutting-edge technique that allows researchers to control the activity of specific neurons with light. By using light-sensitive proteins, researchers can precisely activate or inhibit nAChRs in real-time, enabling the study of how nAChRs contribute to synaptic transmission and behavior in living organisms.
- This technique has been invaluable in studying the role of nAChRs in neural circuits, as it allows for the precise manipulation of receptor activity in specific brain regions. Optogenetic tools have been used to explore the effects of nAChR activation on cognition, addiction, and motor function, offering real-time insights into the dynamic role of these receptors in the brain.

Single-Cell RNA Sequencing

- Single-cell RNA sequencing (scRNA-seq) has opened up new possibilities for understanding gene expression at the single-cell level. This technique allows researchers to identify the specific cells in which nAChRs are expressed, as well as how their expression changes in response to different stimuli or in disease states.
- scRNA-seq has been particularly useful in identifying new subtypes of nAChRs and understanding their differential expression across different regions of the brain. This information can help identify new therapeutic targets and improve our understanding of how nAChRs function in health and disease.

The Role of Animal Models in Advancing nAChR Understanding

Animal models have played a crucial role in advancing our knowledge of nAChRs and their role in various neurological and psychiatric disorders. These models allow for the investigation of nAChR function in a controlled environment and provide a means of testing new therapeutic approaches.

Rodent Models

- Rodents, particularly mice and rats, are the most commonly used animal models in nAChR research. Through genetic manipulation, such as knockout or transgenic models, researchers can study the effects of specific nAChR subunits or the absence of nAChRs on behavior and disease progression.
- For example, knockout mice lacking specific nAChR subunits have been used to explore the roles of different subtypes in cognition, motor function, and addiction. These models have provided critical insights into how changes in nAChR function contribute to diseases like Alzheimer's and Parkinson's.

Non-Human Primates

- Non-human primates, such as monkeys, are another important animal model for studying the role of nAChRs in complex behaviors, including learning, memory, and motor control. These models are particularly valuable in studying higher-order cognitive functions and the effects of nAChR modulation on these processes.
- Primate models are also useful for studying the potential effects of nAChR-targeted therapies in more complex and behaviorally relevant settings, providing a bridge between rodent models and human clinical trials.

Advancements in Imaging

- Recent advancements in imaging techniques, such as positron emission tomography (PET) and magnetic resonance imaging (MRI), have allowed for in vivo imaging of nAChR expression and function in animal models. PET imaging, in particular, enables researchers to visualize the distribution of nAChRs in the brain and assess how their activity is altered in disease states.
- These imaging techniques are paving the way for non-invasive studies of nAChRs in living animals, facilitating the development of new diagnostic tools and therapies.

How nAChRs are Studied In Vitro and In Vivo

In Vitro Studies

- In vitro studies, conducted outside of living organisms in controlled laboratory settings, are essential for understanding the molecular mechanisms underlying nAChR function. Techniques such as patch-clamp electrophysiology, ligand binding assays, and calcium imaging are commonly used to study nAChR activity in cell cultures.
- These methods allow researchers to isolate and manipulate nAChRs at the cellular level, providing insights into how nAChRs function as ion channels, how they respond to different ligands, and how they contribute to synaptic signaling.

In Vivo Studies

- In vivo studies are conducted in living organisms and allow researchers to observe how nAChRs function in the context of complex neural circuits and behavior. These studies often involve the use of animal models to assess the effects of nAChR modulation on behavior, cognition, and disease progression.
- Advanced techniques, such as optogenetics and chemogenetics, allow researchers to selectively activate or inhibit nAChRs in specific brain regions in real-time, providing valuable insights into the role of nAChRs in behavior and disease.

The Future of nAChR Research and Potential Therapies

The future of nAChR research holds exciting possibilities, with the potential to uncover new therapeutic strategies for a wide range of neurological and psychiatric disorders. As our understanding of nAChR function and their role in disease deepens, new drug candidates and therapeutic approaches targeting these receptors are likely to emerge.

Targeted nAChR Modulation

- As the field advances, there is increasing interest in developing drugs that selectively target specific nAChR subtypes. This approach may allow for more precise modulation of nAChR activity, minimizing side effects while enhancing therapeutic efficacy. For example, targeting the α7 nAChR subtype may offer potential benefits in the treatment of Alzheimer's disease, while modulating α4β2 nAChRs could be useful for treating addiction or schizophrenia.
- Advances in structural biology and cryo-electron microscopy (cryo-EM) are enabling researchers to gain detailed insights into the three-dimensional structure of nAChRs. These insights may lead to the development of more potent and selective nAChR modulators.

Gene Therapy

- Gene therapy, which involves the delivery of genetic material to modify nAChR expression or function, is a promising area of research. By restoring or enhancing nAChR function in regions of the brain where receptors are deficient or dysregulated, gene therapy could provide a powerful tool for treating diseases like Parkinson's and Alzheimer's.
- CRISPR-Cas9 technology offers the potential for precise gene editing of nAChR subunits, enabling researchers to study the effects of specific mutations or restore normal receptor function in disease models.

Personalized Medicine

- With advances in genomics and precision medicine, there is growing potential for developing personalized therapies for nAChR-related diseases. By analyzing an individual's genetic profile, it may be possible to tailor nAChR-targeted treatments to their specific needs, improving outcomes and minimizing side effects.
- Pharmacogenomics, the study of how genetic variations influence an individual's response to drugs, could play a key role in the development of personalized nAChR therapies.

Conclusion

The field of nAChR research is advancing rapidly, with the development of cutting-edge technologies like CRISPR, optogenetics, and single-cell RNA sequencing providing new insights into the structure, function, and role of these receptors in health and disease. Animal models continue to play a critical role in our understanding of nAChRs, while in vitro and in vivo studies are shedding light on the mechanisms by which nAChRs influence behavior and disease progression. The future of nAChR research holds immense promise, with targeted therapies, gene editing, and personalized medicine offering new opportunities for treating neurological and psychiatric disorders. As our knowledge of nAChRs deepens, we can expect significant advancements in the development of therapies that target these receptors to improve the lives of individuals affected by a wide range of diseases.

Summary

In this chapter, we have explored the latest advances in nAChR research, including groundbreaking techniques like CRISPR, optogenetics, and single-cell RNA sequencing. These tools are providing new insights into nAChR function and their role in health and disease. The continued use of animal models and in vitro studies is expanding our understanding of these receptors, while promising therapeutic strategies, such as targeted nAChR modulation, gene therapy, and personalized medicine, hold great potential for treating a wide range of neurological and psychiatric disorders. The future of nAChR research is bright, with the potential to develop more effective and precise therapies for these debilitating conditions.

Chapter 11: nAChRs in Disease Mechanisms

Nicotinic acetylcholine receptors (nAChRs) play an integral role in many physiological processes, influencing synaptic transmission, cognition, motor control, and more. Given their widespread involvement in the nervous system, disruptions in nAChR signaling are implicated in a range of diseases, from neurodegenerative disorders to cancers. Understanding how nAChRs contribute to disease mechanisms is crucial for developing therapeutic interventions. This chapter explores the molecular pathways involving nAChRs in disease progression, with a focus on inflammation, oxidative stress, and epigenetic regulation.

Molecular Pathways Involving nAChRs in Disease Progression

The activation of nAChRs triggers a cascade of intracellular signaling events that can have profound effects on neuronal and non-neuronal cells. These signaling pathways are central to the role of nAChRs in disease mechanisms. The main pathways associated with nAChR activation are as follows:

Calcium Ion Influx and Signal Transduction

- Upon binding acetylcholine (ACh) or other agonists, nAChRs open their ion channels, allowing the influx of calcium (Ca^{2+}) ions into the cell. This calcium influx activates various intracellular signaling pathways, including those involving kinases such as protein kinase C (PKC) and phosphoinositide 3-kinase (PI3K). These pathways regulate gene expression, cell survival, and synaptic plasticity.
- Dysregulation of calcium homeostasis has been implicated in many neurodegenerative diseases, including Alzheimer's and Parkinson's. An overactivation of nAChRs can lead to excessive calcium influx, contributing to excitotoxicity and neuronal damage.

Phosphorylation of Receptor Subunits

- The phosphorylation of specific nAChR subunits (e.g., α7 and α4) plays a role in receptor desensitization, trafficking, and cellular responses to external stimuli. Altered phosphorylation patterns can impair receptor function and contribute to disease pathogenesis.
- In Alzheimer's disease (AD), changes in the phosphorylation status of nAChRs, especially the α7 subtype, have been linked to impaired cognitive function and the accumulation of amyloid plaques, a hallmark of the disease.

Increased Reactive Oxygen Species (ROS) Production

- nAChRs are involved in the regulation of redox balance within cells. Activation of nAChRs, particularly in neurons, can stimulate mitochondrial activity and promote the production of reactive oxygen species (ROS), which can lead to oxidative stress when not properly regulated.
- Oxidative stress is a key contributor to neurodegenerative diseases. For instance, in Parkinson's disease (PD), oxidative damage to dopaminergic neurons in the substantia nigra is thought to result from nAChR dysregulation. Similarly, ROS are implicated in the progression of Alzheimer's disease, contributing to neuronal apoptosis and inflammation.

The Relationship Between Inflammation and nAChR Signaling

Inflammation is a common pathological feature in many diseases, including neurodegenerative diseases, autoimmune disorders, and even cancer. nAChRs are recognized for their involvement in modulating immune responses, with recent studies highlighting their role in both the initiation and resolution of inflammation.

nAChRs and the Inflammatory Response

- nAChRs are expressed not only in neurons but also in immune cells such as microglia, macrophages, and T-cells. The activation of these receptors on immune cells can modulate the release of pro-inflammatory cytokines.
- The α7 nAChR, in particular, is involved in the cholinergic anti-inflammatory pathway. This receptor, when activated, can inhibit the release of tumor necrosis factor-alpha (TNF-α) and other pro-inflammatory cytokines, leading to a reduction in inflammation. This pathway has been shown to play a protective role in diseases like rheumatoid arthritis, colitis, and sepsis.

Chronic Inflammation in Neurodegenerative Diseases

- In neurodegenerative diseases, chronic inflammation is thought to contribute to the progression of disease. In Alzheimer's disease, microglial activation leads to the release of pro-inflammatory cytokines, which can worsen neuronal damage.
- The role of nAChRs in modulating microglial activation is critical, as altered nAChR signaling may exacerbate neuroinflammation. For example, reduced expression of α7 nAChRs has been associated with heightened neuroinflammation and the acceleration of disease progression in Alzheimer's.

nAChRs as Therapeutic Targets in Inflammation

- Targeting nAChRs for modulating inflammatory responses has become a promising therapeutic approach. Agonists that selectively activate α7 nAChRs could offer a way to mitigate excessive inflammation without impairing normal immune function.
- The development of α7-selective agonists or partial agonists is underway, with early-stage clinical trials suggesting their potential for treating inflammatory conditions and neurodegenerative diseases.

nAChRs and Oxidative Stress

Oxidative stress is a critical factor in many diseases, as it causes cellular damage through the accumulation of reactive oxygen species (ROS), which overwhelm the antioxidant defenses of cells. nAChRs are implicated in the regulation of ROS, particularly in the central nervous system (CNS).

Role of nAChRs in ROS Production

- Activation of nAChRs on neurons or glial cells can lead to mitochondrial stimulation, increasing ROS production. While low levels of ROS are important for normal cell signaling, excessive ROS can damage cellular components such as lipids, proteins, and DNA.
- In neurodegenerative diseases, including Alzheimer's and Parkinson's, this ROS-mediated damage contributes to the breakdown of neuronal function. In particular, dopaminergic neurons in Parkinson's disease are highly susceptible to oxidative stress, and dysregulated nAChR activity has been implicated in amplifying this oxidative damage.

Antioxidant Effects of nAChRs

- Paradoxically, certain nAChRs, especially the α7 subtype, are involved in cellular antioxidant defense mechanisms. α7 nAChR activation has been shown to stimulate the production of antioxidants and promote cellular repair pathways.
- This dual role of nAChRs in both promoting and counteracting oxidative stress makes them an attractive target for therapeutic interventions aimed at reducing oxidative damage in neurodegenerative diseases.

Epigenetic Regulation of nAChRs in Disease Contexts

Epigenetic modifications—changes in gene expression or cellular phenotype that do not involve alterations to the underlying DNA sequence—are emerging as important regulators of nAChR expression and function in disease contexts. Several mechanisms, including DNA methylation, histone modification, and non-coding RNA regulation, can influence the expression of nAChRs and contribute to disease progression.

DNA Methylation and nAChRs

- DNA methylation, a key epigenetic modification, has been shown to regulate the expression of nAChR genes. In several neurological diseases, such as Alzheimer's disease, abnormal DNA methylation patterns can lead to changes in nAChR expression, particularly with the α7 subtype.
- The downregulation of α7 nAChRs through hypermethylation is associated with increased neuroinflammation and oxidative stress in the brain, contributing to disease progression.

Histone Modifications and nAChR Expression

- Histone modifications, which influence chromatin structure and gene expression, also play a role in nAChR regulation. In the context of neurodegenerative diseases, altered histone acetylation and methylation patterns have been observed, which affect the transcriptional regulation of nAChR genes.
- These modifications may lead to a decrease in nAChR expression, further exacerbating the neurodegenerative process. In Alzheimer's disease, for example, the loss of nAChRs due to epigenetic changes is thought to impair synaptic function and contribute to cognitive decline.

Non-Coding RNAs and nAChR Regulation

- Non-coding RNAs, such as microRNAs, have been shown to regulate nAChR expression. In diseases like Parkinson's and schizophrenia, alterations in the expression of specific microRNAs that target nAChR genes may contribute to the pathophysiology of these disorders.
- For instance, microRNAs that downregulate α4β2 nAChRs have been linked to addiction and cognitive deficits. Understanding these regulatory networks opens new avenues for potential therapeutic interventions.

Conclusion

The relationship between nAChRs and disease mechanisms is multifaceted, involving intricate molecular pathways related to inflammation, oxidative stress, and epigenetic regulation. nAChRs are central to the pathogenesis of various neurological and psychiatric disorders, including Alzheimer's, Parkinson's, and addiction. As research continues to uncover the molecular details of these interactions, nAChRs are emerging as potential therapeutic targets for modulating disease progression. Strategies that aim to enhance or normalize nAChR function, whether through receptor agonists, epigenetic modulation, or inflammation control, hold significant promise for the treatment of a wide range of conditions. Understanding the role of nAChRs in disease will be key to the development of more effective and targeted therapies in the future.

Summary

This chapter explored the complex role of nAChRs in disease mechanisms, focusing on their involvement in molecular pathways related to inflammation, oxidative stress, and epigenetic regulation. Dysregulation of nAChR signaling contributes to the progression of several diseases, including neurodegenerative disorders and cancer. Modulating nAChRs to reduce inflammation, control oxidative stress, and regulate gene expression holds

Chapter 12: nAChRs as Drug Targets

Nicotinic acetylcholine receptors (nAChRs) have long been recognized as critical components in neurotransmission and synaptic plasticity. With their widespread distribution throughout both the central and peripheral nervous systems, nAChRs play pivotal roles in many physiological processes, including cognition, motor function, and immune response. Consequently, nAChRs have emerged as important drug targets for a wide variety of diseases and conditions, from neurodegenerative disorders to addiction, and even cancer. In this chapter, we explore the current state of drug discovery targeting nAChRs, challenges in drug design, and the potential for future therapeutic advancements.

Current Drugs Targeting nAChRs

Several classes of drugs that target nAChRs are currently used in clinical practice or are in development. These drugs can be categorized into agonists, antagonists, and modulators, each playing a distinct role in altering nAChR activity.

nAChR Agonists

- **Nicotine Replacement Therapy (NRT)**: The most well-known agonist targeting nAChRs is nicotine itself. NRT products (e.g., nicotine patches, gums, and lozenges) are used to help smokers quit by alleviating withdrawal symptoms and cravings. Nicotine binds to nAChRs, particularly in the brain, activating them and providing a sense of satisfaction, which helps ease the transition to a smoke-free life.

- **Varenicline (Chantix)**: A partial agonist at α4β2 nAChRs, varenicline is used as a smoking cessation aid. It stimulates nAChRs to release dopamine, reducing cravings and withdrawal symptoms, while also blocking nicotine from activating the receptors, thereby reducing the reinforcing effects of smoking.

- **Galantamine and Donepezil**: These drugs are acetylcholinesterase inhibitors that indirectly increase acetylcholine levels, thus enhancing nAChR activation. They are primarily used in the treatment of Alzheimer's disease to improve cognitive function by compensating for cholinergic deficits.

nAChR Antagonists

- **Curare and Pancuronium**: These drugs are used in clinical settings to induce muscle relaxation during surgery. They act as competitive antagonists at nAChRs in the neuromuscular junction, preventing acetylcholine from binding and inhibiting muscle contraction.
- **α7 nAChR Antagonists**: Research has shown that α7 nAChR antagonists could play a role in treating neuropsychiatric disorders such as schizophrenia and anxiety. By inhibiting overactive α7 nAChRs, these drugs aim to reduce symptoms of these disorders.

nAChR Allosteric Modulators

- **Anatoxins**: Certain natural compounds, such as anatoxin-a, bind to nAChRs allosterically, modulating receptor function without directly competing with acetylcholine. These agents have been studied for their potential use in treating diseases like Alzheimer's, where improving nAChR function may enhance cognition.
- **Positive Allosteric Modulators (PAMs)**: Emerging research is focused on the development of PAMs that selectively enhance the function of nAChRs without overstimulation. These compounds may offer more precise control over nAChR activity, reducing the risk of adverse effects seen with traditional agonists.

Development of Novel nAChR-Targeted Therapies

The discovery of nAChRs' involvement in numerous diseases has spurred the development of novel therapies targeting these receptors. This section highlights the current efforts and strategies being explored to harness nAChRs for therapeutic benefit.

Neurodegenerative Diseases

- **Alzheimer's Disease (AD)**: In AD, the cholinergic system is impaired, leading to cognitive decline. Research has focused on enhancing nAChR activity, particularly α7 and α4β2 subtypes, to restore cholinergic signaling. Drugs that act as agonists at these receptors may help ameliorate cognitive symptoms and slow disease progression. New compounds in preclinical and clinical stages aim to selectively target these receptor subtypes to avoid unwanted side effects.
- **Parkinson's Disease (PD)**: In PD, dopaminergic neurons are lost, but nAChRs in the brain's basal ganglia can modulate dopamine release. Studies have explored using α6β2* nAChR agonists to enhance dopaminergic transmission, offering a potential adjunctive treatment to traditional dopaminergic therapies.

Addiction and Smoking Cessation

The role of nAChRs in nicotine addiction is well-established, but they also play a part in addiction to other substances, including alcohol, cocaine, and opioids. Research into α4β2 and α7 nAChR-targeted therapies is ongoing, with the goal of developing drugs that reduce the rewarding effects of addictive substances. For example, compounds that block or partially activate specific nAChR subtypes may reduce cravings and withdrawal symptoms associated with drug use.

Pain and Inflammation

nAChRs have been implicated in modulating pain perception and inflammation. Studies suggest that α7 nAChRs in immune cells and neurons play a role in reducing inflammation, while their activation may also modulate nociceptive pathways. Drugs that target these receptors, particularly in conditions like rheumatoid arthritis, inflammatory bowel disease, and chronic pain, are being investigated for their potential to reduce inflammation and provide pain relief.

Cancer

nAChRs have been shown to influence cancer cell proliferation, migration, and metastasis. Specifically, α7 nAChRs are expressed in several tumor types, including non-small cell lung cancer and breast cancer, and activation of these receptors can promote tumor growth. Targeting nAChRs in cancer cells has the potential to inhibit tumor progression. Research into nAChR antagonists, particularly α7 antagonists, is underway, with promising preclinical results suggesting that blocking these receptors could serve as an effective cancer therapy.

Challenges in Drug Design and Delivery

Despite the potential for nAChR-targeted therapies, several challenges complicate the development of safe and effective drugs:

Receptor Subtype Selectivity

- One of the biggest hurdles in nAChR drug design is achieving subtype selectivity. There are multiple subtypes of nAChRs, each with different distributions and functions. For example, α7 nAChRs are involved in cognitive processes and inflammation, while α4β2 nAChRs play a key role in addiction. Developing drugs that selectively target one subtype without affecting others is difficult but essential to minimizing side effects.
- Current research is exploring novel allosteric modulators and biased ligands that can selectively activate specific receptor subtypes, potentially leading to more targeted therapies with fewer adverse effects.

Blood-Brain Barrier (BBB) Penetration

Many of the conditions treated with nAChR-targeted therapies are neurological in nature, requiring drugs that can cross the blood-brain barrier (BBB). The design of molecules that are both nAChR-selective and able to penetrate the BBB is a significant challenge. Strategies such as nanoparticle drug delivery, prodrug design, and the development of small molecule agonists or antagonists are being pursued to overcome this barrier.

Toxicity and Safety

- While nAChR-targeted drugs offer therapeutic potential, their use can lead to adverse effects, such as excessive stimulation (e.g., seizures, neurotoxicity) or receptor desensitization (leading to tolerance or withdrawal symptoms). A delicate balance must be struck between therapeutic benefit and safety. Moreover, long-term use of certain nAChR drugs may lead to unintended consequences, such as changes in receptor expression or function that could exacerbate disease.
- Rigorous clinical trials and post-marketing surveillance will be necessary to fully understand the safety profiles of new nAChR-targeted therapies.

Clinical Trials and Future Directions

The next frontier in nAChR-targeted therapies lies in clinical translation. Numerous drugs targeting nAChRs are currently undergoing clinical trials across a range of conditions. Promising results from early-phase trials in Alzheimer's disease, Parkinson's disease, and addiction may soon lead to new treatment options for patients.

Future research will likely focus on refining the specificity and efficacy of these therapies. The use of advanced technologies such as CRISPR-Cas9 to manipulate gene expression in animal models, optogenetics to control receptor activation in real-time, and artificial intelligence (AI) to predict receptor-ligand interactions will accelerate the development of new drugs.

Additionally, personalized medicine, guided by genetic and epigenetic profiling, may help optimize treatment plans by identifying patients who are most likely to benefit from nAChR-targeted therapies based on their receptor profiles.

Conclusion

nAChRs represent an exciting and versatile class of drug targets, with applications ranging from neurodegenerative diseases and addiction to cancer and inflammation. Current therapies are already having an impact on smoking cessation, cognitive disorders, and pain management, but there is still much to learn about the full potential of nAChR modulation. With continued advancements in drug design, clinical testing, and the integration of cutting-edge technologies, nAChR-targeted therapies hold promise for treating a wide variety of conditions and improving patient outcomes in the years to come.

Chapter 13: Genetic Mutations and nAChR Dysfunction

Nicotinic acetylcholine receptors (nAChRs) are integral to a wide range of physiological processes, from cognitive function and muscle contraction to immune system regulation. Their dysfunction is implicated in various neurological disorders, diseases, and conditions. Genetic mutations that affect nAChRs can lead to altered receptor function, which in turn can contribute to disease susceptibility, altered drug responses, and even congenital disorders. This chapter explores the genetic basis of nAChR dysfunction, the impact of mutations on receptor function and disease mechanisms, and the emerging role of personalized medicine in addressing these challenges.

Genetic Mutations Leading to nAChR Dysfunction

The nAChR family is composed of various subtypes, each encoded by distinct genes. Mutations in these genes can lead to changes in the receptor structure, function, or expression. These mutations can occur in both coding regions (resulting in protein changes) or non-coding regions (influencing gene expression levels or receptor trafficking). The consequences of such mutations are wide-ranging, from mild to severe impairments in synaptic transmission and receptor activity.

Subunit Gene Mutations

- **α4 and β2 Subunits**: Mutations in the genes encoding the α4 and β2 subunits of nAChRs are the most common genetic causes of congenital myasthenic syndromes (CMS). These mutations can alter receptor function by reducing the number of functional receptors at the neuromuscular junction, impairing neuromuscular transmission and leading to muscle weakness. For example, mutations that reduce receptor affinity for acetylcholine or impair ion channel opening can disrupt signal transmission between motor neurons and muscles.
- **α7 Subunit Mutations**: The α7 subunit of nAChRs is critical for cognitive function and has been implicated in several neurological disorders. Mutations in the gene encoding α7 nAChRs (CHRNA7) are associated with conditions like schizophrenia, Alzheimer's disease, and epilepsy. These mutations can result in receptor desensitization or altered channel conductance, affecting cognitive processing and synaptic plasticity. Furthermore, α7 nAChRs are involved in the regulation of inflammatory pathways, and mutations may contribute to immune dysregulation in autoimmune diseases.
- **Other Subunits**: Mutations in other nAChR subunits, such as α3, α5, and β4, have been associated with various neurological conditions, including Parkinson's disease, epilepsy, and neurodegeneration. Disruptions in the expression or functionality of these subunits can lead to impaired neurotransmission,

Effects on Receptor Trafficking and Expression mood.

Certain mutations can affect the trafficking of nAChRs to the cell surface, limiting the number of functional receptors available for synaptic transmission. This phenomenon is particularly relevant in diseases such as congenital myasthenic syndrome, where nAChR trafficking defects result in a reduced number of receptors at the neuromuscular junction. Receptor density and distribution are essential for proper synaptic signaling, and disruptions in this process can lead to severe impairments in neuromuscular and neuronal communication.

Impact on Disease Susceptibility and Drug Response

Genetic mutations that affect nAChR function can influence an individual's susceptibility to a variety of diseases, as well as their response to drug treatments. Understanding the genetic basis of nAChR dysfunction is critical for developing targeted therapies and improving personalized medicine approaches.

Neurological Diseases

- **Schizophrenia**: The α7 nAChR has been implicated in the pathophysiology of schizophrenia. Genetic variants that result in reduced α7 nAChR function are associated with cognitive deficits in individuals with schizophrenia. Research has shown that individuals with these genetic mutations may have a diminished response to certain cognitive-enhancing therapies. Targeting the α7 nAChR with selective agonists is a promising therapeutic approach, but the variability in receptor function due to genetic differences must be taken into account in treatment planning.
- **Alzheimer's Disease**: In Alzheimer's disease, nAChRs, particularly α7 nAChRs, play a role in modulating cognitive function and reducing neuroinflammation. Mutations in the CHRNA7 gene may impair receptor function and contribute to the cognitive decline observed in Alzheimer's patients. Personalized medicine approaches, including genetic testing, can help identify patients with specific α7 nAChR mutations who may benefit from drugs designed to enhance receptor activity or protect against neuroinflammation.

- **Parkinson's Disease**: α4β2 nAChRs are involved in dopaminergic signaling in the brain, and mutations in these receptors can influence susceptibility to Parkinson's disease. Studies have shown that individuals with certain mutations in the α4 or β2 subunits may experience altered dopamine release or signaling, which could contribute to motor deficits. Identifying individuals at genetic risk for Parkinson's disease through nAChR mutation analysis could enable early intervention strategies aimed at preserving dopaminergic function.

Drug Response Variability

- Genetic mutations in nAChRs can also impact an individual's response to drugs that target these receptors. For example, patients with mutations in the α4β2 subunit of the receptor may have a different response to nicotine replacement therapies, which rely on nAChR activation to alleviate withdrawal symptoms. Similarly, individuals with α7 nAChR mutations may respond differently to drugs designed to enhance cognitive function in conditions like Alzheimer's or schizophrenia.
- Personalized medicine, guided by genetic testing for nAChR mutations, holds the potential to optimize drug efficacy and minimize side effects. By tailoring treatments to an individual's genetic profile, clinicians can choose the most appropriate therapeutic approach based on how their nAChRs are likely to respond to specific drugs.

Role in Congenital Disorders (e.g., Congenital Myasthenic Syndromes)

Congenital myasthenic syndromes (CMS) are a group of inherited disorders caused by mutations in genes encoding nAChRs or other components of the neuromuscular junction. These syndromes result in impaired neuromuscular transmission and are characterized by muscle weakness, fatigue, and exercise intolerance. Mutations in the nAChR subunits ($\alpha 1$, $\beta 1$, δ, and ε) are the primary causes of CMS, with the most common mutations being in the $\alpha 1$ and $\beta 1$ subunits.

Types of CMS

- **Synaptic CMS**: Mutations affecting the $\alpha 1$ subunit of the nAChR lead to reduced receptor function and impaired synaptic transmission. These mutations are typically associated with a milder form of CMS, which presents with muscle weakness that worsens with exercise but improves with rest.
- **Endplate CMS**: Mutations in the $\beta 1$ subunit or in proteins involved in receptor clustering at the neuromuscular junction (e.g., rapsyn) can cause more severe forms of CMS. These mutations result in a marked reduction in receptor density at the neuromuscular junction, leading to profound muscle weakness and respiratory difficulties.

The identification of specific genetic mutations in CMS has improved the diagnosis and treatment of these conditions. Genetic screening for CMS-causing mutations enables clinicians to tailor treatment approaches, such as the use of acetylcholinesterase inhibitors or other supportive therapies, to the patient's specific genetic profile.

Personalized Medicine and Genetic Screening for nAChRs

The advent of personalized medicine has opened new avenues for the treatment of diseases related to nAChR dysfunction. Genetic screening for nAChR mutations can help identify individuals at risk for disorders like schizophrenia, Alzheimer's, and congenital myasthenic syndromes. By understanding an individual's genetic makeup, clinicians can predict disease risk, tailor therapeutic interventions, and monitor drug responses more effectively.

Genetic Testing

- Advances in genomic sequencing have made genetic testing more accessible and affordable. Whole-exome sequencing (WES) and whole-genome sequencing (WGS) are increasingly used to identify mutations in nAChR subunit genes. These tools can provide a comprehensive view of the genetic landscape, revealing potential mutations that may contribute to nAChR-related disorders.
- In clinical practice, genetic tests for nAChR-related mutations are becoming more common, particularly in the context of neuromuscular disorders. For example, genetic screening for CMS can confirm a diagnosis and guide treatment decisions, while testing for α7 nAChR mutations in patients with schizophrenia or cognitive impairment may help identify individuals who could benefit from specific cognitive-enhancing therapies.

Targeted Therapies

Personalized medicine based on genetic profiles may allow for the development of targeted therapies that specifically address the underlying genetic mutations. For example, in patients with CMS caused by α1 nAChR mutations, drugs that enhance receptor function or compensate for defective receptors may be developed to improve neuromuscular transmission and alleviate symptoms.

Conclusion

Genetic mutations affecting nAChRs play a crucial role in the pathophysiology of a range of neurological and neuromuscular disorders. Understanding the molecular basis of nAChR dysfunction is key to improving diagnostic accuracy, disease management, and drug development. As genetic screening becomes more widespread and personalized medicine advances, clinicians will be better equipped to treat nAChR-related diseases with precision, ultimately improving patient outcomes. Future research into the genetic underpinnings of nAChR dysfunction, combined with novel therapeutic strategies, holds great promise for the treatment of these challenging conditions.

Chapter 14: The Role of nAChRs in Aging

As we age, the body's physiological systems undergo significant changes, impacting overall health and functionality. One area of profound change is the nervous system, where alterations in neurotransmitter systems, particularly the cholinergic system, can lead to cognitive decline, motor dysfunction, and a range of age-associated neurodegenerative diseases. Among the key players in this system are nicotinic acetylcholine receptors (nAChRs), which are essential for cognitive processes like learning and memory, motor control, and synaptic plasticity. This chapter explores the role of nAChRs in the aging process, how aging affects their expression and function, and strategies to modulate nAChRs to mitigate age-related cognitive decline and enhance overall quality of life.

nAChRs and Age-Related Cognitive Decline

Cognitive decline is one of the most significant and distressing aspects of aging. Conditions like Alzheimer's disease (AD) and other forms of dementia are characterized by progressive loss of memory, attention, and executive function. A major contributing factor to cognitive decline in aging is the dysfunction of the cholinergic system, which involves acetylcholine (ACh) and its receptors, including nAChRs. These receptors are critically involved in the processes that underlie learning, memory, and synaptic plasticity, making them essential in maintaining cognitive function throughout life.

Cholinergic Hypothesis of Aging and Dementia

- According to the cholinergic hypothesis of aging and Alzheimer's disease, a decline in acetylcholine (ACh) production and nAChR function is closely linked to cognitive impairments. Studies have shown that individuals with Alzheimer's disease have a significant reduction in nAChRs, particularly those in the hippocampus and cortex, regions involved in learning and memory. This reduction in receptor activity impairs neurotransmission and synaptic plasticity, key mechanisms for cognitive function.
- The loss of α7 and α4β2 nAChR subtypes is particularly noteworthy, as these subtypes are involved in modulating neurotransmitter release and maintaining synaptic plasticity. In aging, the upregulation of certain pro-inflammatory cytokines, which often occur in neurodegenerative diseases, can also lead to a reduction in nAChR expression and function, contributing to a feedback loop that exacerbates cognitive decline.

Synaptic Plasticity and Memory Formation

nAChRs play a key role in synaptic plasticity, which is essential for memory formation and the adaptability of neural circuits. In older adults, the reduced function of these receptors impairs the plasticity of neural connections, leading to difficulties in learning new information or retaining memories. The α7 nAChRs are particularly important in long-term potentiation (LTP), a process that strengthens synaptic connections in response to stimuli and underlies memory consolidation. Reduced α7 nAChR function with age impairs LTP, contributing to memory deficits in elderly individuals.

Changes in nAChR Expression with Age

The expression and functionality of nAChRs change over the lifespan, with substantial decreases observed in older adults. These changes have significant implications for the cholinergic system's ability to support cognitive and motor function.

Age-Dependent Decline in nAChR Density

- Studies have shown that the density of nAChRs decreases in various brain regions as part of the natural aging process. This decline is particularly evident in the hippocampus, cortex, and striatum, regions that are central to cognitive and motor control. The reduction in receptor density is often accompanied by a decline in the levels of acetylcholine, leading to decreased activation of nAChRs.
- This age-related loss of receptor function contributes to impaired cognitive performance, including difficulties with attention, memory recall, and executive function. The exact mechanisms behind this decline are still being studied but may involve factors such as reduced gene expression, receptor desensitization, and changes in receptor trafficking.

Changes in Receptor Subtypes

- Different nAChR subtypes show varying patterns of expression across the lifespan. For example, α7 nAChRs, which are involved in cognitive processes and neuroprotection, tend to be particularly vulnerable to age-related decline. α4β2 nAChRs, involved in both cognitive and motor functions, also decrease with age, particularly in individuals with neurodegenerative conditions such as Parkinson's and Alzheimer's diseases.
- The decrease in α7 nAChRs may contribute not only to cognitive decline but also to increased neuroinflammation, as these receptors are involved in the regulation of the immune response in the brain. The decline in these receptors may lead to an increase in pro-inflammatory cytokines, which further damage neural tissue and accelerate neurodegeneration.

The Effects of Aging on the Cholinergic System

Aging not only leads to a decline in nAChR density and function but also affects the broader cholinergic system, including the synthesis, release, and breakdown of acetylcholine (ACh).

Reduced Acetylcholine Synthesis

As we age, there is a decrease in the activity of choline acetyltransferase (ChAT), the enzyme responsible for synthesizing acetylcholine. This leads to lower ACh levels in the brain, particularly in regions critical for cognition, such as the hippocampus and cortex. Reduced ACh availability can lead to impaired signaling at nAChRs, further exacerbating cognitive decline.

Increased AChE Activity

The enzyme acetylcholinesterase (AChE) breaks down acetylcholine after it has been released into the synaptic cleft. With age, AChE activity tends to increase, leading to more rapid degradation of ACh and further reducing its availability for nAChR activation. This imbalance between ACh synthesis and breakdown contributes to the dysfunction of the cholinergic system in aging.

Increased Inflammation

Age is also associated with increased inflammation in the brain, often referred to as "neuroinflammation." This inflammatory response can impair the function of nAChRs by altering receptor expression and promoting oxidative stress. Chronic inflammation may reduce the availability of ACh receptors and impair neurotransmission, accelerating age-related cognitive decline.

Strategies to Modulate nAChRs in Aging Populations

Given the pivotal role of nAChRs in age-related cognitive decline, there is growing interest in developing strategies to modulate nAChRs to preserve or enhance cognitive function in older adults. These strategies can range from pharmacological interventions to lifestyle modifications that promote receptor function and health.

Pharmacological Modulation

- **Cholinergic Drugs**: Drugs that enhance cholinergic activity, such as acetylcholinesterase inhibitors (e.g., donepezil, rivastigmine), are commonly used to treat Alzheimer's disease. These drugs increase the availability of acetylcholine at the synapse, indirectly enhancing nAChR activation. While these treatments can provide some symptomatic relief, they do not stop the progression of neurodegeneration.
- **Selective nAChR Agonists**: There is growing interest in the development of drugs that directly target specific nAChR subtypes. For example, selective agonists for α7 nAChRs have shown promise in preclinical models of Alzheimer's disease, enhancing cognitive function and reducing neuroinflammation. Such drugs could offer more targeted treatment options for age-related cognitive decline.

Lifestyle Modifications

- **Exercise**: Regular physical activity has been shown to enhance cognitive function in aging populations by promoting neurogenesis, increasing ACh release, and upregulating nAChR expression. Exercise may help to counteract the age-related decline in receptor density and function, preserving cognitive performance.
- **Diet**: Nutrients that support brain health, such as omega-3 fatty acids, antioxidants, and B vitamins, can help maintain nAChR function and reduce neuroinflammation. Additionally, reducing exposure to environmental toxins and smoking cessation may prevent further damage to the cholinergic system.
- **Cognitive Training**: Engaging in mentally stimulating activities, such as learning new skills, solving puzzles, or practicing memory exercises, can enhance synaptic plasticity and improve cognitive reserve. This may help to offset the functional decline of nAChRs in aging brains.

Conclusion

Nicotinic acetylcholine receptors (nAChRs) are central to the maintenance of cognitive function and neural health, and their decline with age is closely linked to age-related cognitive impairments and neurodegenerative diseases. Understanding the role of nAChRs in aging, as well as the factors that contribute to their dysfunction, is essential for developing targeted therapeutic strategies. By modulating nAChR activity through pharmacological interventions, lifestyle changes, or a combination of both, it may be possible to slow the cognitive decline associated with aging and improve the quality of life for older adults. As research in this area continues to evolve, new breakthroughs in nAChR-targeted therapies hold promise for addressing the growing challenges of aging populations worldwide.

4o mini

You said:

write chapter 15 for the book Mastering Nicotinic Acetylcholine Receptors (nAChRs) Chapter 1: Introduction to Nicotinic Acetylcholine Receptors History of acetylcholine discovery Role of acetylcholine in the nervous system Overview of receptor types: muscarinic vs.

nicotinic Structure and function of nAChRs Chapter 2: Molecular Biology of nAChRs Genetic encoding of nAChRs Gene families involved in nAChR synthesis mRNA and protein expression of nAChRs Localization in different tissues Chapter 3: Biochemistry of Nicotinic Receptors Receptor subtypes and their significance Ligand binding and receptor activation Ion channel function Desensitization and adaptation mechanisms Chapter 4: Physiology of nAChRs Role of nAChRs in synaptic transmission Functions in the central and peripheral nervous systems The connection between nAChRs and motor function Interaction with other neurotransmitter systems Chapter 5: Pharmacology of nAChRs Overview of nAChR agonists and antagonists Synthetic and natural compounds affecting nAChRs Therapeutic and toxicological implications The role of nAChRs in addiction and dependence Chapter 6: nAChRs in the Central Nervous System Function in cognitive processes: learning and memory Involvement in neurodegenerative diseases: Alzheimer's, Parkinson's Effects on mood, anxiety, and depression Research on nAChRs in neuropsychological disorders Chapter 7: nAChRs in the Peripheral Nervous System Autonomic nervous system regulation Muscular function and nAChRs in motor neurons The role in neuromuscular diseases like myasthenia gravis nAChRs and pain modulation Chapter 8: nAChRs and Addiction Nicotine addiction: Mechanisms and pathways The role of nAChRs in drug abuse (cocaine, alcohol, etc.

) Therapeutic strategies for addiction management Research on potential treatments targeting nAChRs Chapter 9: The Role of nAChRs in Neurological Disorders Alzheimer's disease: Cholinergic dysfunction and nAChRs Parkinson's disease: Dopamine and nAChR interactions Huntington's disease and the cholinergic system Schizophrenia and the role of nAChRs in psychosis Chapter 10: Advances in nAChR Research Cutting-edge techniques in nAChR research: CRISPR, optogenetics, etc. The role of animal models in advancing nAChR understanding How nAChRs are studied in vitro and in vivo The future of nAChR research and potential therapies Chapter 11: nAChRs in Disease Mechanisms Molecular pathways involving nAChRs in disease progression The relationship between inflammation and nAChR signaling nAChRs and oxidative stress Epigenetic regulation of nAChRs in disease contexts Chapter 12: nAChRs as Drug Targets Current drugs targeting nAChRs Development of novel nAChR-targeted therapies Challenges in drug design and delivery Clinical trials and future directions Chapter 13: Genetic Mutations and nAChR Dysfunction Genetic mutations leading to nAChR dysfunction Impact on disease susceptibility and drug response Role in congenital disorders (e.g.

, congenital myasthenic syndromes) Personalized medicine and genetic screening for nAChRs Chapter 14: The Role of nAChRs in Aging nAChRs and age-related cognitive decline Changes in nAChR expression with age The effects of aging on the cholinergic system Strategies to modulate nAChRs in aging populations Chapter 15: Diagnostic Approaches Involving nAChRs Techniques for measuring nAChR activity and expression Imaging methods: PET scans, MRIs in nAChR studies Biomarkers for diseases involving nAChRs Genetic tests for nAChR-related disorders Chapter 16: nAChRs in Cancer The role of nAChRs in tumor progression nAChR signaling pathways in cancer cells Potential for using nAChRs as biomarkers in oncology Targeting nAChRs for cancer treatment Chapter 17: Clinical Applications of nAChR Modulators Current clinical applications of nAChR agonists and antagonists Use in neurological diseases (e.g.

, Alzheimer's, Parkinson's) Emerging treatments for mental health disorders The role of nAChRs in pain management and inflammation Chapter 18: Toxicology of nAChRs The toxic effects of nicotine and other nAChR ligands Understanding nAChR-related side effects in drug development The balance between therapeutic and toxic effects Case studies of nAChR toxicity in clinical practice Chapter 19: Modulation of nAChRs in Neuroprotection How nAChRs contribute to neuronal survival The potential for neuroprotective drugs targeting nAChRs nAChRs in the context of stroke and neurodegeneration Current and future neuroprotective therapies Chapter 20: Ethical Considerations in nAChR Research Ethical issues in addiction research Ethical challenges in genetic testing for nAChR-related conditions The use of animal models in nAChR studies Balancing scientific progress with ethical responsibility Chapter 21: Future of nAChR Targeted Therapies Trends in drug development for nAChR-related diseases The role of AI and machine learning in nAChR research Global collaboration in nAChR research and therapy development Next-generation therapies and precision medicine Chapter 22: nAChRs in Non-Neuronal Tissues Role in immune cells and inflammation nAChRs in epithelial and endothelial cells The contribution of nAChRs to organ function beyond the nervous system Implications for non-neuronal therapies Chapter 23: Environmental and Lifestyle Factors Affecting nAChRs The effects of diet, exercise, and stress on nAChR expression Environmental toxins and their impact on nAChR function The role of smoking and other lifestyle factors in nAChR-related diseases Public health perspectives on nicotine use and nAChR modulation Chapter 24:

Interdisciplinary Approaches to nAChR Research Collaborations across pharmacology, neuroscience, and genetics The importance of bioinformatics in

studying nAChRs Insights from systems biology in understanding nAChR functions Integrating nAChR research with other fields like immunology and cancer research Chapter 25: Conclusion and Future Directions Recap of key takeaways from the book The potential for future breakthroughs in nAChR research How understanding nAChRs can shape medical practices in the future Final thoughts on advancing therapeutic strategies involving nAChRs

ChatGPT said:

ChatGPT

Chapter 15: Diagnostic Approaches Involving nAChRs

As research into nicotinic acetylcholine receptors (nAChRs) progresses, so too does the development of innovative diagnostic techniques to assess their function and role in health and disease. The ability to measure the activity, expression, and localization of nAChRs in vivo and in vitro is critical for advancing our understanding of these receptors and their involvement in various conditions, from neurodegenerative diseases to cancer and addiction. In this chapter, we will explore the current diagnostic approaches used to study nAChRs, including imaging technologies, biomarkers, genetic testing, and other diagnostic methods.

Techniques for Measuring nAChR Activity and Expression

Several methods are employed to measure nAChR activity and expression at both the cellular and systemic levels. These techniques provide valuable insight into the physiological and pathological roles of nAChRs.

Radioligand Binding Assays

- Radioligand binding assays have long been used to measure the density and distribution of nAChRs. These assays use radiolabeled ligands that specifically bind to nAChRs, allowing researchers to quantify receptor density in various tissues. Radioligand binding studies have been instrumental in identifying changes in nAChR expression in conditions like Alzheimer's disease, Parkinson's disease, and nicotine addiction.
- Commonly used ligands include [125I] α-bungarotoxin, which binds to α7 nAChRs, and [3H]nicotine, which targets both α4β2 and α7 subtypes. These techniques can be applied to brain tissue, peripheral tissues, and even post-mortem samples to assess receptor function and distribution.

Electrophysiological Methods

- Patch-clamp electrophysiology is a powerful technique used to measure the ion channel activity of nAChRs. This method allows for the precise measurement of current flow through nAChRs in isolated cells or brain slices, providing insights into receptor kinetics, ligand binding, and ion conductance.
- Electrophysiological methods can also be used to study the effects of potential drugs or modulators on nAChR activity. For example, studying the response of nAChRs to nicotine or synthetic agonists can help identify their role in diseases such as addiction or neurodegeneration.

Fluorescence-Based Imaging

- Fluorescence-based imaging techniques have become increasingly important for studying nAChR function in living organisms. Fluorescently labeled ligands or antibodies can be used to track nAChR expression in real time, allowing researchers to monitor receptor dynamics in response to pharmacological interventions or disease states.
- Techniques such as single-molecule fluorescence resonance energy transfer (smFRET) and total internal reflection fluorescence (TIRF) microscopy can provide detailed spatial and temporal information about nAChR trafficking, clustering, and activation at the synapse. These methods are particularly useful for studying receptor behavior in neuronal and non-neuronal cells.

Mass Spectrometry and Proteomics

- Mass spectrometry and proteomic techniques can be used to study the expression of nAChRs at the protein level. By analyzing protein extracts from tissues or cells, researchers can identify the specific subtypes of nAChRs present and determine how their expression is altered in disease states.
- Proteomics also allows for the identification of nAChR-associated proteins, which may help uncover the molecular pathways and signaling networks that nAChRs participate in. This is particularly useful for exploring how nAChRs contribute to neurological diseases like Alzheimer's and Parkinson's, where receptor expression and signaling are often disrupted.

Imaging Methods: PET Scans, MRIs, and nAChR Studies

Imaging technologies, particularly positron emission tomography (PET) and magnetic resonance imaging (MRI), offer non-invasive ways to study nAChRs in living organisms, including human patients. These methods provide valuable insights into receptor distribution, functionality, and changes over time.

Positron Emission Tomography (PET)

- PET imaging is one of the most powerful non-invasive techniques for studying the distribution of nAChRs in the living brain. PET scans use radiolabeled ligands that specifically bind to nAChRs, allowing for the visualization of receptor densities in different regions of the brain. This has proven invaluable in both basic research and clinical diagnostics.
- PET imaging has been particularly useful in studying neurodegenerative diseases. For example, in Alzheimer's disease, PET scans using α4β2 selective radioligands can help assess the loss of nAChRs in key regions like the hippocampus and cortex, which are involved in memory and cognitive function.
- Additionally, PET imaging is used to evaluate the effects of drugs targeting nAChRs, providing critical data for clinical trials testing nAChR agonists or antagonists in conditions such as addiction and dementia.

Magnetic Resonance Imaging (MRI)

- While MRI is primarily used for structural imaging, recent advancements have allowed MRI techniques to be adapted for functional studies of nAChRs. Functional MRI (fMRI) can be used to detect changes in brain activity associated with nAChR function, although it does not directly measure receptor activity.
- By coupling MRI with PET, or using MRI-guided spectroscopy, researchers can gain a more comprehensive understanding of the relationship between receptor density, brain activity, and disease states. This multimodal imaging approach holds great potential for diagnosing conditions linked to nAChR dysfunction, such as Parkinson's disease or schizophrenia.

Biomarkers for Diseases Involving nAChRs

Biomarkers are crucial for diagnosing diseases and assessing the efficacy of therapeutic interventions. In the context of nAChRs, biomarkers can be used to identify receptor dysfunction, monitor disease progression, and predict treatment outcomes.

Biomarkers of nAChR Dysfunction

- Alterations in nAChR expression and activity have been linked to several neurological and psychiatric conditions, including Alzheimer's disease, Parkinson's disease, schizophrenia, and addiction. Specific biomarkers associated with nAChRs may include altered levels of receptor subtypes, changes in acetylcholine (ACh) metabolism, and variations in associated proteins involved in nAChR trafficking and signaling.
- For example, reduced α4β2 nAChR availability is commonly observed in Alzheimer's disease, and this can be measured using PET scans with radiolabeled nicotine or other selective ligands. Similarly, changes in α7 nAChR expression can serve as a biomarker for conditions like schizophrenia and depression.

Peripheral Biomarkers

- While brain imaging provides critical insights into central nAChR activity, peripheral biomarkers, such as changes in blood or cerebrospinal fluid (CSF), can also provide valuable information. For instance, elevated levels of certain cytokines or neuroinflammatory markers may indicate altered nAChR function in the central nervous system.
- In the context of smoking-related diseases, peripheral biomarkers related to nAChR activation (such as nicotine metabolites) are commonly used to assess nicotine exposure and addiction status. These biomarkers can be employed to monitor the success of smoking cessation treatments and predict relapse.

Genetic Tests for nAChR-Related Disorders

Genetic testing for nAChR-related disorders is becoming increasingly important as the role of genetic mutations in nAChR dysfunction becomes clearer. Genetic variations in nAChR subunits can lead to altered receptor function, which may predispose individuals to diseases or influence their response to drugs.

Mutations in nAChR Subunits

- Genetic mutations in nAChR subunits, such as those in the CHRNA4 gene (encoding the α4 subunit) and the CHRNB2 gene (encoding the β2 subunit), have been linked to several disorders, including congenital myasthenic syndromes and nicotine addiction. Genetic testing can identify individuals carrying these mutations, which may help in early diagnosis and personalized treatment.
- Whole-genome sequencing and targeted gene panels are powerful tools for detecting nAChR-related mutations. By identifying these mutations, clinicians can assess disease risk, guide treatment decisions, and predict drug responses in patients with neurodegenerative diseases or addiction.

Pharmacogenomics

Pharmacogenomic testing can also be used to determine how an individual's genetic makeup affects their response to nAChR-targeted therapies. For example, genetic variations in nAChR genes may influence the efficacy of nicotine replacement therapies or other drugs targeting nAChRs. This approach can help tailor treatments to maximize effectiveness and minimize adverse effects.

Conclusion

The diagnostic approaches to studying nicotinic acetylcholine receptors (nAChRs) have advanced significantly, with a wide range of techniques now available to measure receptor activity, expression, and dysfunction. From radioligand binding assays and electrophysiological methods to non-invasive imaging techniques like PET and MRI, these tools provide crucial insights into the role of nAChRs in health and disease. Additionally, the use of biomarkers and genetic testing allows for early diagnosis, personalized treatment, and monitoring of disease progression. As our understanding of nAChRs continues to evolve, these diagnostic methods will play an increasingly important role in advancing both basic research and clinical practice, particularly in the fields of neurodegenerative diseases, addiction, and cancer.

Chapter 16: nAChRs in Cancer

Nicotinic acetylcholine receptors (nAChRs), primarily known for their roles in the nervous system, have garnered increasing attention in cancer research due to their involvement in various processes critical to tumor progression, metastasis, and resistance to treatment. While nAChRs are most commonly associated with neuronal signaling, emerging evidence suggests they play a significant role in the regulation of cancer cell growth, survival, migration, and invasion. This chapter explores the role of nAChRs in cancer biology, including their signaling pathways, potential as biomarkers, and the possibility of targeting them for cancer treatment.

The Role of nAChRs in Tumor Progression

Cancer cells, including those from lung, breast, prostate, and colon cancers, have been shown to express nAChRs, with certain subtypes more prevalently involved in tumorigenesis. In the tumor microenvironment, nAChRs can influence various cellular processes such as proliferation, survival, angiogenesis, and immune evasion.

Cell Proliferation and Survival

- nAChRs are involved in the regulation of cell proliferation through activation of intracellular signaling cascades that promote cell cycle progression. The binding of acetylcholine or nicotine to nAChRs on cancer cells can activate downstream signaling pathways such as the mitogen-activated protein kinase (MAPK) and phosphoinositide 3-kinase (PI3K)/Akt pathways. These pathways are key regulators of cell survival and proliferation.
- Additionally, nAChR activation may enhance the expression of anti-apoptotic proteins, such as Bcl-2, contributing to resistance against programmed cell death. This allows tumor cells to survive and proliferate under conditions where normal cells would undergo apoptosis.

Angiogenesis and Metastasis

- Angiogenesis, the formation of new blood vessels, is critical for tumor growth and metastasis. nAChR signaling has been shown to promote angiogenesis by stimulating the release of pro-angiogenic factors like vascular endothelial growth factor (VEGF). The activation of nAChRs on endothelial cells can trigger the MAPK and PI3K/Akt pathways, leading to increased VEGF production and enhanced angiogenesis. This creates a supportive microenvironment for tumor growth.
- nAChRs also play a role in the metastasis of cancer cells. By activating signaling pathways that regulate cell migration, nAChRs facilitate the invasion of cancer cells into surrounding tissues and their spread to distant organs. Studies have demonstrated that nicotine, acting through nAChRs, can promote the epithelial-mesenchymal transition (EMT), a process critical for the invasion of cancer cells and metastasis.

Immune Evasion

- Cancer cells can manipulate immune responses to evade detection and destruction by the host immune system. nAChRs expressed on immune cells, such as T lymphocytes and macrophages, influence immune responses within the tumor microenvironment. By modulating immune cell function, nAChRs can contribute to immune suppression, promoting a microenvironment conducive to tumor growth.
- For instance, activation of nAChRs on tumor-associated macrophages (TAMs) can lead to the release of immunosuppressive cytokines, thereby inhibiting the activity of anti-tumor immune cells. Similarly, nAChR activation on dendritic cells may impair their ability to present antigens effectively, further facilitating immune evasion.

nAChR Signaling Pathways in Cancer Cells

The signaling pathways activated by nAChRs in cancer cells are complex and involve a variety of intracellular messengers that regulate multiple cellular processes. While the specific pathways can vary depending on the cancer type and receptor subtype, common mechanisms include the following:

MAPK/ERK Pathway

The MAPK pathway is one of the most well-established signaling pathways activated by nAChR stimulation. This pathway is involved in regulating cell growth, differentiation, and survival. In cancer cells, nAChR activation leads to the phosphorylation of extracellular signal-regulated kinase (ERK), a key component of the MAPK pathway, which subsequently promotes cell proliferation and survival.

PI3K/Akt Pathway

The PI3K/Akt pathway is another critical signaling cascade modulated by nAChRs. This pathway regulates various cellular functions, including metabolism, growth, survival, and motility. Activation of nAChRs on cancer cells can lead to the activation of PI3K, which in turn phosphorylates and activates Akt. Akt signaling promotes cell survival by inhibiting apoptosis and stimulating cell cycle progression.

JAK/STAT Pathway

In some cancers, nAChR signaling has been linked to the Janus kinase (JAK)/signal transducer and activator of transcription (STAT) pathway. Activation of this pathway by nAChRs can result in the upregulation of genes involved in cell proliferation, survival, and immune evasion. For example, STAT3 activation is associated with increased production of pro-survival cytokines and resistance to chemotherapy.

NF-κB Pathway

The nuclear factor kappa-light-chain-enhancer of activated B cells (NF-κB) pathway, which regulates inflammation and immune responses, is also influenced by nAChR signaling in cancer cells. The activation of this pathway promotes the transcription of genes involved in cell survival, inflammation, and metastasis. Nicotine-induced activation of nAChRs can lead to the activation of NF-κB in tumor cells, which contributes to tumor growth and resistance to treatment.

Potential for Using nAChRs as Biomarkers in Oncology

Given their involvement in various stages of tumor progression, nAChRs have the potential to serve as valuable biomarkers in cancer. The expression patterns of specific nAChR subtypes, such as α7, α4β2, and β2, have been linked to cancer diagnosis, prognosis, and therapeutic response.

Diagnostic Biomarkers

The presence of nAChRs on cancer cells can serve as a diagnostic biomarker for certain types of tumors. For example, elevated expression of α7 nAChRs has been found in lung cancer cells, and their detection may help in identifying patients with early-stage lung cancer. PET imaging using radiolabeled ligands targeting nAChRs has shown promise in detecting tumors based on their nAChR expression profile.

Prognostic Biomarkers

The level of nAChR expression in tumors may also correlate with prognosis. High expression of nAChRs, particularly the α7 subtype, has been associated with more aggressive tumor behavior and poor prognosis in cancers such as breast and prostate cancer. Monitoring nAChR expression could, therefore, help predict disease outcomes and guide treatment strategies.

Predictive Biomarkers

nAChRs may also act as predictive biomarkers for response to specific therapies. For example, patients with tumors expressing high levels of nAChRs may benefit from treatments that target these receptors, such as nAChR antagonists. Conversely, tumors with low or absent nAChR expression may be less responsive to such therapies.

Targeting nAChRs for Cancer Treatment

The ability to target nAChRs for cancer treatment holds significant promise. Several strategies are being explored to modulate nAChR activity and exploit their role in cancer progression.

nAChR Antagonists

Antagonists that block nAChRs, particularly the α7 subtype, may inhibit cancer cell proliferation, migration, and survival. For example, α7 nAChR antagonists have shown anti-tumor activity in preclinical models of lung cancer and other malignancies. By blocking nAChR-mediated signaling, these drugs could reduce tumor growth and enhance the efficacy of chemotherapy and immunotherapy.

Targeting nAChRs with Small Molecule Inhibitors

Small molecule inhibitors that specifically target nAChRs are another promising therapeutic approach. These inhibitors could block the signaling pathways activated by nAChRs, such as MAPK and PI3K/Akt, which are critical for cancer cell survival. In particular, targeting nAChRs in combination with other therapeutic modalities, such as chemotherapy or radiation therapy, may improve treatment outcomes by preventing cancer cell resistance.

Immunotherapy and nAChR Modulation

Recent studies have also explored the potential of combining nAChR-targeted therapies with immunotherapies. By modulating the immune response through nAChRs on immune cells, it may be possible to enhance the efficacy of immune checkpoint inhibitors or adoptive T cell therapies. This dual approach could improve cancer treatment by both targeting tumor cells directly and boosting the body's immune response.

Gene Therapy and CRISPR/Cas9

Gene-editing technologies like CRISPR/Cas9 hold potential for targeting specific nAChR subtypes or modulating their expression in cancer cells. By knocking down or editing genes that encode nAChRs, researchers could potentially reduce tumor growth and metastasis. This approach could offer a more precise and long-lasting method of targeting nAChRs in cancer therapy.

Conclusion

The role of nicotinic acetylcholine receptors in cancer is a rapidly growing area of research. nAChRs contribute to various aspects of tumor progression, including cell proliferation, survival, migration, and immune evasion. As a result, nAChRs have emerged as promising biomarkers for cancer diagnosis, prognosis, and treatment response. Furthermore, targeting nAChRs with specific antagonists, small molecule inhibitors, and gene therapies holds great potential for improving cancer treatment outcomes. As our understanding of nAChRs in cancer deepens, we may unlock new avenues for therapeutic intervention, offering hope for patients suffering from a wide range of malignancies.

Chapter 18: Toxicology of nAChRs

Nicotinic acetylcholine receptors (nAChRs) are central to many physiological processes, especially in the nervous system, where they mediate synaptic transmission and neuronal communication. However, when misregulated or over-activated, these receptors can also contribute to toxicity and adverse effects. This chapter delves into the toxicological aspects of nAChRs, exploring the mechanisms through which their dysfunction leads to harmful effects, the role of nAChRs in toxicological side effects, and how this knowledge can guide safer drug development. We will also look at case studies that highlight the importance of understanding nAChR toxicity in clinical settings.

The Toxic Effects of Nicotine and Other nAChR Ligands

Nicotine, the most well-known ligand for nAChRs, has a significant impact on human health, with both therapeutic and harmful effects. The toxicological consequences of nicotine exposure are broad, ranging from addiction and cardiovascular effects to its role in cancer promotion.

Nicotine Addiction

Nicotine exerts its addictive effects by binding to nAChRs, particularly in the brain's reward centers such as the ventral tegmental area (VTA) and nucleus accumbens. This binding stimulates the release of dopamine, a neurotransmitter associated with pleasure and reinforcement. Chronic activation of nAChRs in these areas leads to dependence and addiction. Understanding the underlying mechanisms of nicotine addiction at the receptor level is crucial for developing effective treatments for nicotine dependence.

Cardiovascular Toxicity

The chronic activation of nAChRs, particularly in the autonomic nervous system, is linked to various cardiovascular risks, including hypertension, arrhythmias, and increased heart rate. Nicotine increases the release of catecholamines (e.g., norepinephrine), which raises blood pressure and induces vasoconstriction. Over time, these effects contribute to the development of cardiovascular diseases, including coronary artery disease and stroke.

Cancer Promotion

While nicotine itself is not a direct carcinogen, its action on nAChRs can contribute to the promotion of cancer. Studies have shown that nicotine and other nAChR ligands can stimulate tumor growth and metastasis by activating signaling pathways involved in cell proliferation, survival, and angiogenesis. Nicotine-induced activation of nAChRs can enhance the growth of existing tumors, particularly in the lungs, and promote resistance to chemotherapy.

Neurotoxic Effects

Prolonged exposure to nicotine and other nAChR agonists can have neurotoxic effects. Chronic nicotine use has been associated with cognitive deficits, including memory impairment and attention disorders. These effects may be linked to dysregulation of nAChRs in the brain, particularly in regions such as the hippocampus and prefrontal cortex, which are involved in learning and memory. Additionally, prolonged nicotine exposure can lead to the desensitization of nAChRs, disrupting normal neurotransmission and cognitive function.

Understanding nAChR-Related Side Effects in Drug Development

The pharmacological properties of nAChRs make them an attractive target for drug development, particularly for diseases like Alzheimer's, Parkinson's, and nicotine addiction. However, as with any receptor system, targeting nAChRs carries the risk of unwanted side effects. These side effects are often linked to the modulation of nAChR subtypes in tissues beyond the central nervous system (CNS), such as the peripheral nervous system and other non-neuronal tissues.

Off-Target Effects

One of the key challenges in nAChR drug development is the receptor's widespread distribution in the body. nAChRs are found not only in the brain but also in the peripheral nervous system, skeletal muscles, and other tissues. Drugs designed to target nAChRs in the CNS can unintentionally affect peripheral tissues, leading to off-target effects such as muscle weakness, gastrointestinal disturbances, and cardiovascular issues. For example, non-selective nAChR agonists can cause excessive stimulation of muscle nAChRs, resulting in neuromuscular toxicity.

Desensitization and Tolerance

Chronic exposure to nAChR agonists, such as nicotine, can lead to receptor desensitization. This means that the receptors become less responsive to stimulation, requiring higher doses to achieve the same effect. This desensitization is a key factor in the development of drug tolerance and dependence. It also complicates the therapeutic use of nAChR modulators, as the body may adapt to the drug's presence, diminishing its efficacy over time.

Toxicity in Clinical Practice

The development of drugs targeting nAChRs requires careful attention to dosing, receptor selectivity, and potential long-term effects. For instance, drugs used to treat neurodegenerative diseases that target nAChRs, such as cholinesterase inhibitors, must be carefully dosed to avoid overstimulation of nAChRs, which can lead to side effects like muscle cramps, dizziness, and nausea. Similarly, the use of nAChR antagonists in the treatment of nicotine addiction needs to be balanced to avoid exacerbating withdrawal symptoms or triggering cardiovascular events.

The Balance Between Therapeutic and Toxic Effects

The therapeutic benefits of nAChR modulators must be weighed against their potential toxicological risks. Striking the right balance between efficacy and safety is a core challenge in drug development. Some nAChR-targeted therapies have shown promise in treating neurological conditions like Alzheimer's and Parkinson's disease, while others are primarily used for smoking cessation. However, these therapies must be used with caution due to the risk of adverse effects.

Selective Targeting

One promising strategy to minimize toxicity while maximizing therapeutic benefit is selective targeting of specific nAChR subtypes. For example, selective α4β2 or α7 nAChR agonists have been investigated for their potential in treating cognitive dysfunctions, such as those observed in Alzheimer's disease. By selectively targeting the receptors that are most relevant to the condition while avoiding those involved in peripheral side effects, it may be possible to reduce toxicity and improve patient outcomes.

Personalized Medicine

Genetic variations in nAChRs, such as polymorphisms in the genes encoding different subtypes, can influence an individual's response to nAChR-targeted drugs. Personalized medicine approaches that take into account a patient's genetic profile could help identify those who are at higher risk for toxicity or who are more likely to benefit from certain treatments. For instance, patients with certain genetic variants of the α7 nAChR may respond differently to drugs targeting this receptor, and adjusting treatment plans accordingly could improve safety and efficacy.

Dose Optimization

Proper dose optimization is crucial to minimize the risk of toxicity. Many nAChR-targeted drugs, such as nicotine replacement therapies, need to be carefully titrated to avoid excessive stimulation of nAChRs. The risk of overdose and toxicity increases when the drugs are taken inappropriately, such as in the case of excessive nicotine use through smoking or e-cigarettes. Similarly, drugs designed to treat conditions like myasthenia gravis need to ensure that nAChR stimulation is sufficient to alleviate symptoms without causing overstimulation of the muscle or nervous system.

Case Studies of nAChR Toxicity in Clinical Practice

Case studies can provide valuable insights into the real-world implications of nAChR toxicity in clinical settings. These examples underscore the importance of understanding the toxic effects of nAChRs and the need for careful management when using drugs that target these receptors.

Nicotine Poisoning

One well-documented case of nAChR toxicity is nicotine poisoning, which occurs when nicotine is absorbed in excessive amounts, either through smoking, e-cigarette use, or nicotine replacement therapies. Symptoms of nicotine poisoning include nausea, vomiting, dizziness, rapid heart rate, and in severe cases, seizures or respiratory failure. Understanding the toxicological effects of nicotine and how it interacts with nAChRs in both the central and peripheral nervous systems is crucial for preventing and managing these cases.

Cholinergic Crisis

A cholinergic crisis occurs when there is an overactivation of cholinergic receptors, including nAChRs, often as a result of excessive acetylcholine or nAChR agonists. This can happen in conditions such as organophosphate poisoning, where inhibitors of acetylcholinesterase lead to prolonged stimulation of nAChRs. Symptoms include muscle weakness, respiratory distress, and paralysis. Proper management of cholinergic crisis involves the use of nAChR antagonists and other supportive therapies to counteract the toxic effects of overactivation.

Myasthenia Gravis and nAChR Dysfunction

Myasthenia gravis, a neuromuscular disorder caused by the autoimmune attack on nAChRs at the neuromuscular junction, serves as another case study. Patients with myasthenia gravis experience muscle weakness, fatigue, and respiratory issues due to impaired transmission at the neuromuscular junction. While treatments like cholinesterase inhibitors help to enhance nAChR function, careful dosing is critical to avoid excessive stimulation, which can lead to toxicity and exacerbate symptoms.

Conclusion

Understanding the toxicology of nicotinic acetylcholine receptors (nAChRs) is essential for advancing both therapeutic strategies and clinical safety. While nAChRs play critical roles in normal physiological processes, their dysregulation or overstimulation can lead to significant toxic effects, including addiction, cardiovascular risks, neurotoxicity, and cancer promotion. Striking a balance between therapeutic efficacy and toxicity remains a challenge in drug development, but with targeted therapies, personalized medicine, and dose optimization, we can mitigate these risks. Future research into the toxicological mechanisms of nAChRs will further illuminate their role in disease and guide the safe use of nAChR-targeted treatments.

Chapter 19: Modulation of nAChRs in Neuroprotection

Nicotinic acetylcholine receptors (nAChRs) have long been recognized for their critical role in neuronal communication and synaptic transmission, but their potential as modulators of neuroprotection is a burgeoning area of research. Beyond their fundamental functions in cognition, motor control, and autonomic regulation, nAChRs have emerged as powerful contributors to neuronal survival, resilience, and repair. This chapter explores how the modulation of nAChRs can be leveraged to protect neurons from injury and degeneration, with a particular focus on their role in stroke, neurodegenerative diseases, and potential therapeutic strategies in neuroprotection.

How nAChRs Contribute to Neuronal Survival

Nicotinic receptors, particularly those of the α7 and α4β2 subtypes, have been shown to play significant roles in maintaining neuronal health and function. These receptors are involved in a variety of intracellular signaling pathways that promote cellular survival, anti-apoptotic mechanisms, and the maintenance of synaptic plasticity.

Activation of Neuroprotective Pathways

- When nAChRs are activated by ligands, they initiate a cascade of intracellular signaling events. The α7-nAChR subtype, for instance, is closely associated with anti-inflammatory effects and neuroprotection. Activation of α7-nAChRs has been shown to reduce the release of pro-inflammatory cytokines and to inhibit microglial activation, which is a key event in the pathology of neurodegenerative diseases and acute brain injury.
- Additionally, the activation of nAChRs can enhance the synthesis of neurotrophic factors, such as brain-derived neurotrophic factor (BDNF) and nerve growth factor (NGF), which are crucial for neuronal survival and regeneration. These factors promote cell growth, survival, and differentiation in response to injury.

Calcium Influx and Survival Signaling

The activation of nAChRs leads to ion flux, particularly calcium ions, into the cell. While excessive calcium influx can be toxic and lead to excitotoxicity, moderate and controlled calcium entry through nAChRs has been shown to trigger protective signaling pathways. For example, the calcium influx through α7-nAChRs activates calmodulin-dependent protein kinase (CaMKII), which then activates the phosphatidylinositol-3-kinase (PI3K)/Akt pathway, a well-known survival pathway that helps to protect neurons from apoptosis.

Inhibition of Apoptosis

nAChR activation has been linked to the suppression of apoptotic signaling. By modulating pathways such as the Bcl-2 family of proteins, nAChRs can reduce cell death in response to various stressors, including oxidative damage and excitotoxicity. This apoptosis-inhibitory effect makes nAChRs a promising target in the context of neurodegeneration and stroke, where cell death is a primary concern.

nAChRs in the Context of Stroke and Neurodegeneration

Stroke and neurodegenerative diseases like Alzheimer's, Parkinson's, and Huntington's disease represent major global health challenges, and the ability to protect neurons from the damage they experience during these conditions is a key focus of modern neuropharmacology. nAChR modulation holds great promise as a therapeutic approach for these diseases.

nAChRs in Stroke

- Stroke, particularly ischemic stroke, results from a blockage of blood flow to the brain, leading to oxygen and nutrient deprivation. This deprives neurons of vital resources and triggers a cascade of events leading to neuronal injury and death. nAChRs, especially those in the α7 subtype, are implicated in mitigating these events.
- Activation of α7-nAChRs in the context of stroke has been shown to attenuate the inflammatory response, which is a critical factor in secondary brain injury following ischemia. Microglial cells, when activated, release inflammatory cytokines that exacerbate neuronal damage. The anti-inflammatory effects of nAChRs may reduce the extent of this damage, leading to improved outcomes post-stroke.

nAChRs in Alzheimer's Disease

- In Alzheimer's disease, one of the key pathological features is the loss of cholinergic neurons, which leads to cognitive decline and memory loss. The α4β2-nAChRs are particularly important in modulating cognitive function and memory processes.
- Studies have shown that targeting these receptors with selective agonists or enhancing their activity can improve cognitive function in animal models of Alzheimer's. Additionally, nAChR-mediated signaling can protect neurons from amyloid-beta-induced toxicity, a hallmark of Alzheimer's pathology. Thus, nAChRs represent a potential avenue for slowing or halting cognitive decline in Alzheimer's patients.

nAChRs in Parkinson's Disease

- Parkinson's disease is characterized by the degeneration of dopaminergic neurons in the substantia nigra, leading to motor impairments such as tremors, rigidity, and bradykinesia. nAChRs, particularly the α6β2* subtype, play a role in modulating dopaminergic transmission, and their dysfunction may contribute to the motor deficits seen in Parkinson's disease.
- Preclinical studies have suggested that enhancing nAChR signaling can restore dopaminergic function and improve motor symptoms. Furthermore, nAChR activation can promote neuroprotection by reducing oxidative stress and neuroinflammation, which are critical contributors to the neurodegenerative process in Parkinson's.

nAChRs in Huntington's Disease

- Huntington's disease, another neurodegenerative disorder, involves the progressive degeneration of neurons in the striatum and cortex. Here, nAChRs also play a neuroprotective role by modulating neurotransmitter release and by participating in the reduction of neuroinflammation.
- Research has shown that nAChR agonists may reduce the neurotoxic effects of glutamate, which contributes to excitotoxicity and neuronal death in Huntington's disease. Therefore, modulating nAChR function in Huntington's could be a promising therapeutic approach to slow disease progression.

Current and Future Neuroprotective Therapies

The growing understanding of nAChRs' role in neuroprotection has led to the development of several therapeutic strategies aimed at harnessing their potential in protecting neurons from injury and degeneration.

Selective nAChR Agonists

- Developing drugs that selectively activate specific nAChR subtypes, such as the α7-nAChR or α4β2-nAChR, holds great promise for neuroprotection. α7-nAChR agonists, in particular, have been shown to exert anti-inflammatory effects and protect neurons from excitotoxicity in models of stroke and neurodegeneration.
- Clinical trials investigating α7-nAChR agonists for Alzheimer's and Parkinson's diseases are ongoing, with promising results suggesting that these drugs can improve cognitive function, reduce neuroinflammation, and protect against neuronal damage.

Cholinesterase Inhibitors

- In Alzheimer's disease, cholinesterase inhibitors (such as donepezil and rivastigmine) are used to increase acetylcholine levels, thereby indirectly enhancing nAChR activation. While these drugs are primarily symptomatic, they may have neuroprotective effects by promoting cholinergic signaling in the brain.
- The combination of cholinesterase inhibitors with nAChR-targeted therapies is an area of active research, as it may offer a synergistic approach to enhance cognitive function and protect against neuronal damage.

Gene Therapy and nAChR Modulation

- Advances in gene therapy may also offer novel ways to enhance nAChR function and protect against neurodegeneration. For example, viral vectors could be used to deliver genes encoding specific nAChR subtypes to regions of the brain that are particularly vulnerable to neurodegeneration, such as the hippocampus in Alzheimer's disease or the substantia nigra in Parkinson's.
- Additionally, gene-editing technologies like CRISPR-Cas9 could be employed to correct genetic mutations that affect nAChR function, potentially providing a long-term solution to diseases like congenital myasthenic syndrome, which is caused by mutations in the nAChR gene.

Combination Therapies

Neuroprotective strategies may be most effective when nAChR modulation is combined with other therapeutic approaches, such as anti-inflammatory agents, antioxidants, or neurotrophic factor-based therapies. The synergy between nAChR modulators and other pharmacological agents could provide a multifaceted approach to protecting neurons from injury and slowing the progression of neurodegenerative diseases.

Conclusion

The modulation of nicotinic acetylcholine receptors (nAChRs) offers a promising frontier in neuroprotection, with the potential to slow or even halt the progression of neurodegenerative diseases and minimize the damage caused by acute events like stroke. Through selective activation of specific nAChR subtypes, enhancement of neuronal survival pathways, and inhibition of neuroinflammation, nAChRs provide a powerful mechanism for protecting the brain from injury. With continued research and the development of targeted therapies, nAChRs may play a central role in the future of neuroprotection, offering new hope for patients suffering from debilitating neurological disorders.

Chapter 20: Ethical Considerations in nAChR Research

Nicotinic acetylcholine receptors (nAChRs) have become central to research in neuroscience, pharmacology, and therapeutic development, with significant progress being made in their role in cognition, addiction, neuroprotection, and even cancer treatment. However, as with any area of advanced biological and pharmacological research, nAChR studies and therapies bring forward numerous ethical considerations. These ethical challenges must be addressed to ensure that progress in nAChR science is achieved responsibly, with attention to societal, individual, and environmental implications.

This chapter explores the ethical issues surrounding the research and clinical applications of nAChRs, including concerns related to addiction, genetic testing, animal experimentation, and the balance between scientific innovation and moral responsibility.

1. Ethical Issues in Addiction Research

One of the most prominent ethical challenges in nAChR research is the study and treatment of nicotine addiction. Nicotine, the primary ligand for nAChRs, is highly addictive, and its abuse is a leading cause of preventable death worldwide. Understanding the pathways that mediate nicotine addiction through nAChRs is essential for developing effective therapies for tobacco addiction and other substance use disorders. However, there are several ethical concerns related to the study of addiction and nicotine dependence.

Informed Consent in Addiction Research

- Research on addiction often involves individuals with substance use disorders, making informed consent a critical issue. Ensuring that participants fully understand the nature of the research, the risks involved, and the potential consequences of participation is crucial. In addiction studies, this becomes especially important when individuals are actively struggling with their dependencies, potentially compromising their decision-making capacity.
- Additionally, research involving nicotine administration or other addictive substances raises questions about the potential harm caused to participants. Ethical guidelines must ensure that the study design minimizes risks while providing valuable scientific data.

The Role of nAChRs in Nicotine Dependence

While studying nAChRs to understand addiction pathways, there is a risk that the research could inadvertently contribute to the promotion of addictive behaviors, particularly if nAChR-targeting drugs or therapies are used to reinforce smoking behaviors. Researchers and clinicians must ensure that such treatments are framed within a broader public health strategy aimed at harm reduction, rather than inadvertently encouraging nicotine dependence.

Therapeutic Use vs. Abuse

The line between using nAChR-based treatments for therapeutic purposes and the risk of developing new dependencies must be carefully managed. While nAChR agonists, such as varenicline, are used to help individuals quit smoking, their use could raise concerns about the potential for misuse or dependency in vulnerable populations. Balancing the therapeutic benefits of nAChR modulation with the risk of abuse is a major ethical concern that must be carefully monitored in clinical practice.

2. Ethical Challenges in Genetic Testing for nAChR-Related Conditions

Genetic research on nAChRs has revealed a wealth of information about how genetic mutations in nAChR subunits may lead to diseases such as congenital myasthenic syndromes, neurodegenerative diseases, and even certain cancers. However, as with all genetic research, ethical challenges arise, particularly with regard to privacy, consent, and the potential misuse of genetic information.

Genetic Privacy and Discrimination

As genetic screening for nAChR-related mutations becomes more prevalent, concerns about genetic privacy and discrimination emerge. The possibility of individuals being denied employment or insurance based on their genetic predisposition to nAChR-related diseases is a real concern. Genetic information must be handled with the utmost care to protect individuals' rights and prevent the misuse of sensitive data.

Informed Consent and Genetic Counseling

In the context of genetic testing for nAChR-related conditions, informed consent becomes more complex. Individuals undergoing genetic testing must fully understand the potential implications of learning about their genetic makeup, including the risks of discovering predispositions to serious health conditions. Genetic counseling is essential to ensure that patients are equipped with the information they need to make informed decisions about testing and subsequent treatment options.

Ethical Implications of Genetic Editing

The ability to edit genes, including those encoding nAChR subunits, raises profound ethical questions. Technologies such as CRISPR-Cas9 have made gene editing more accessible, but the potential for altering the genetic code of individuals – especially germline editing – poses significant ethical dilemmas. The long-term consequences of editing genes involved in nAChR function, particularly in the context of preventing diseases like myasthenic syndrome or neurodegeneration, must be carefully weighed against the potential for unintended side effects or the alteration of human genetics in ways that may affect future generations.

Personalized Medicine and the Risk of Inequality

As genetic research continues to shape personalized medicine, the potential for unequal access to genetic tests and tailored nAChR-based therapies must be considered. Socioeconomic factors, access to healthcare, and education can all influence an individual's ability to benefit from these innovations. Efforts should be made to ensure that these advances in genetic medicine do not exacerbate existing health disparities but instead promote equity in healthcare access.

3. The Use of Animal Models in nAChR Studies

Animal models have been invaluable in advancing our understanding of nAChR function and its role in various diseases. However, the ethical concerns surrounding the use of animals in research are substantial, and the research community must continuously assess whether the potential benefits of animal studies outweigh the ethical costs.

Animal Welfare and Research Necessity

- The primary ethical concern in using animals in nAChR research is the potential for harm and suffering. Procedures that involve invasive interventions, such as the implantation of electrodes or the administration of harmful substances, can cause distress to the animals involved. Researchers are obligated to ensure that any pain or distress is minimized, and that animals are treated humanely throughout the research process.
- Ethical review boards play a vital role in assessing whether the research is justified and whether alternatives, such as non-animal models or advanced in vitro techniques, could achieve similar results without involving animals. The 3Rs principle — Replacement, Reduction, and Refinement — should be central to any nAChR-related animal research protocol.

Justification for Animal Use

Given the complexity of nAChR function and its diverse roles across various systems, animal models are often essential for studying the full range of effects and interactions of nAChRs in living organisms. However, each study must be critically evaluated for its scientific merit and potential to advance knowledge that directly benefits human health. The ethical justification of animal research hinges on the significance of the knowledge gained and the necessity of using animals in that specific context.

Public Perception and Ethical Oversight

The use of animals in nAChR research also raises concerns among the public and advocacy groups, which can lead to calls for stricter oversight or the reduction of animal-based research. Transparency in research practices, as well as the clear communication of the ethical frameworks guiding these studies, is crucial to maintaining public trust and acceptance of scientific research involving animals.

4. Balancing Scientific Progress with Ethical Responsibility

As nAChR research progresses, particularly in areas like gene editing, addiction treatment, and neuroprotection, it is essential to balance scientific advancements with ethical considerations. The pursuit of new knowledge and therapies must not come at the cost of human dignity, animal welfare, or societal well-being.

Collaborative Ethical Guidelines

- Given the far-reaching implications of nAChR research, ethical guidelines should be established collaboratively, involving scientists, ethicists, policymakers, and public health experts. These guidelines must address issues such as patient consent, genetic privacy, animal welfare, and the equitable distribution of new therapies.
- Ethical frameworks should evolve in parallel with technological advancements to ensure that research is conducted responsibly, with appropriate safeguards in place.

Global Cooperation and Ethical Standards

- nAChR research spans the globe, with scientists from diverse cultures and regulatory environments contributing to the field. It is important that international ethical standards are developed and followed to ensure consistency and fairness in the treatment of human subjects, animals, and the environment.
- Global cooperation in setting ethical guidelines can also help mitigate the risks of unethical practices, such as genetic manipulation or exploitation of vulnerable populations for research purposes.

Conclusion

The ethical considerations surrounding nAChR research are multifaceted and require careful attention to ensure that scientific progress is pursued responsibly. While the potential benefits of nAChR modulation in treating addiction, neurological disorders, and other diseases are vast, they must be tempered by a commitment to the well-being of individuals, animals, and society. By prioritizing informed consent, genetic privacy, animal welfare, and global ethical standards, nAChR research can continue to advance in a manner that reflects both scientific and moral responsibility.

Chapter 21: Future of nAChR Targeted Therapies

Nicotinic acetylcholine receptors (nAChRs) have been at the forefront of medical and pharmacological research for decades, owing to their profound role in the nervous system, as well as their involvement in a broad range of diseases. The study of nAChRs has not only enhanced our understanding of fundamental neuroscience but also paved the way for innovative therapeutic strategies. As our knowledge of nAChR biology advances, so too does the potential for targeted therapies that could transform the management of diseases such as Alzheimer's, Parkinson's, cancer, addiction, and more.

This chapter examines the future of nAChR-targeted therapies, focusing on emerging trends in drug development, the application of artificial intelligence (AI) and machine learning, global collaboration in research, and the promise of next-generation therapies and precision medicine.

1. Trends in Drug Development for nAChR-Related Diseases

The therapeutic targeting of nAChRs has become a rapidly evolving area of drug development, with several promising molecules and strategies already in clinical trials. Understanding the molecular and physiological mechanisms by which nAChRs contribute to disease has opened new avenues for pharmacological intervention. However, the path forward for nAChR-based therapies is complex, with several key trends likely to shape their development.

Selective Modulation of Receptor Subtypes

One of the most promising approaches in nAChR drug development is the selective targeting of specific nAChR subtypes. This allows for more precise modulation of receptor activity, which could minimize off-target effects and enhance therapeutic efficacy. For example, α7 nAChRs, involved in cognitive processes, are potential targets for Alzheimer's disease treatments, while α4β2 nAChRs are implicated in addiction and smoking cessation therapies. The growing understanding of nAChR diversity and its tissue-specific functions will enable the development of drugs that target only the relevant subtypes for each disease.

Allosteric Modulation

Allosteric modulators of nAChRs offer a promising strategy for therapeutic intervention. These compounds bind to sites on the receptor distinct from the acetylcholine binding site, either enhancing or inhibiting the receptor's response to the neurotransmitter. This approach provides greater control over receptor activity and could lead to fewer side effects compared to traditional agonists or antagonists. The development of selective allosteric modulators for specific nAChR subtypes is a key area of interest for treating neurodegenerative diseases, addiction, and mood disorders.

Gene Therapy and Genetic Modulation

With the advent of gene therapy technologies, there is growing interest in using genetic approaches to modulate nAChR function in specific tissues. This could involve the delivery of genetic material to either upregulate or downregulate the expression of certain nAChR subunits, thereby restoring normal receptor function in diseases where nAChRs are dysfunctional. Techniques such as CRISPR/Cas9 offer the potential for precise gene editing that could correct genetic mutations in nAChRs associated with diseases like congenital myasthenic syndrome or neurodegenerative disorders.

Repurposing Existing Drugs

In addition to developing novel drugs, the repurposing of existing pharmaceuticals could provide a faster and more cost-effective route for nAChR-targeted therapies. For example, drugs already approved for other indications might show promise in modulating nAChRs for conditions like Alzheimer's, Parkinson's, or addiction. Exploring this repurposing strategy can accelerate the introduction of nAChR-based treatments to the market.

2. The Role of AI and Machine Learning in nAChR Research

Artificial intelligence (AI) and machine learning (ML) are transforming biomedical research, and their application to nAChR studies holds tremendous promise. AI and ML algorithms can process vast amounts of data to uncover complex patterns and predict outcomes that would be difficult for traditional methods to discern. In the context of nAChR research, these technologies can accelerate drug discovery, personalize treatment strategies, and enhance our understanding of receptor function.

Drug Discovery and Development

- AI and ML have already begun to play a pivotal role in drug discovery, including the identification of novel nAChR ligands. By analyzing large datasets of chemical compounds and their interactions with nAChRs, AI can predict which molecules are most likely to bind effectively to specific nAChR subtypes. This can significantly shorten the drug development timeline and reduce the need for trial-and-error experimentation.
- Furthermore, AI algorithms can optimize drug designs by predicting the pharmacokinetics and toxicity profiles of potential compounds, increasing the likelihood of success in clinical trials.

Personalized Medicine

AI and ML also hold the potential to revolutionize personalized medicine for diseases related to nAChR dysfunction. By analyzing genetic data, biomarkers, and clinical histories, AI can help identify the most effective nAChR-targeted therapies for individual patients. For example, individuals with certain genetic variations in nAChR subunits may respond better to specific treatments, and AI can aid in tailoring therapies to these unique profiles. This approach could improve treatment outcomes and reduce adverse effects.

Predicting Disease Progression and Treatment Response

AI models can be used to predict disease progression and patient responses to nAChR-targeted therapies. By integrating data from clinical trials, imaging, and molecular biology, machine learning algorithms can provide insights into how diseases like Alzheimer's or Parkinson's progress and which nAChR interventions are most likely to be beneficial at different stages. This predictive capability can help clinicians make more informed decisions about treatment regimens and improve patient care.

3. Global Collaboration in nAChR Research and Therapy Development

The future of nAChR-targeted therapies is intrinsically linked to global collaboration in research. The complexity of nAChR biology and its involvement in multiple systems across the body requires expertise from a range of disciplines, including neuroscience, pharmacology, genetics, and bioinformatics. By fostering international partnerships, researchers can share resources, data, and insights that will accelerate the development of effective therapies.

Collaborative Networks and Consortia

- Multinational research consortia focused on nAChR research can pool knowledge and resources, enabling more robust and comprehensive studies. These collaborations can include pharmaceutical companies, academic institutions, and government agencies, working together to address the multifaceted challenges associated with nAChR-related diseases.
- For example, international collaborations between neuroscientists and pharmacologists could advance our understanding of how nAChRs contribute to diseases like Alzheimer's, facilitating the development of novel therapeutic approaches that can be tested across diverse populations.

Data Sharing and Open Access

The rise of open-access platforms for scientific data has already begun to transform the way researchers collaborate. By sharing genomic, proteomic, and clinical data related to nAChRs, researchers can accelerate discoveries and identify new therapeutic targets. Open-source databases and repositories of nAChR-related studies will continue to enhance global efforts to find solutions for diseases like Parkinson's and addiction.

Public and Private Sector Partnerships

Collaboration between the public and private sectors is critical for translating research into clinical applications. Government agencies and nonprofit organizations can provide funding and infrastructure for large-scale nAChR studies, while private pharmaceutical companies bring the necessary expertise in drug development and regulatory approval. By working together, these sectors can fast-track the development of nAChR-targeted therapies and ensure that promising discoveries reach the clinic more quickly.

4. Next-Generation Therapies and Precision Medicine

The ultimate goal of nAChR-targeted therapies is to provide highly effective treatments with minimal side effects for diseases ranging from neurodegeneration to addiction. Next-generation therapies, including gene therapies, biologics, and precision medicine approaches, promise to revolutionize the way we treat nAChR-related conditions.

Gene Editing and Delivery Systems

Advances in gene editing technologies, particularly CRISPR/Cas9, hold great promise for correcting genetic mutations in nAChR genes that contribute to diseases. For example, in conditions like congenital myasthenic syndromes, where nAChR mutations impair neuromuscular transmission, gene therapy could restore normal function. The challenge lies in developing safe and effective delivery methods to introduce gene therapies into specific tissues, particularly the brain.

Biologics and Antibody-Based Therapies

Biologics, including monoclonal antibodies that target nAChRs, could be developed as precision therapies for diseases like Alzheimer's and cancer. These therapies could target specific nAChR subtypes involved in disease processes, offering a higher degree of specificity and efficacy than traditional small-molecule drugs.

Nanotechnology in nAChR Therapy Delivery

Nanotechnology offers innovative solutions for the targeted delivery of nAChR modulators. Nanoparticles or nanocarriers could be designed to deliver drugs directly to the site of action, minimizing off-target effects and enhancing therapeutic efficacy. This could be particularly important for treating diseases that affect the brain, where drugs need to cross the blood-brain barrier to be effective.

Regenerative Medicine

Regenerative medicine approaches, including stem cell therapy, may offer the potential to repair or replace damaged tissues that involve nAChR dysfunction. By using stem cells to generate new neurons or repair damaged neuromuscular junctions, it may be possible to restore normal nAChR function and reverse disease progression in some cases.

Conclusion

The future of nAChR-targeted therapies is incredibly promising, with numerous innovative strategies poised to transform the way we treat a wide range of diseases. From selective receptor modulation and gene editing to AI-driven drug discovery and personalized medicine

Chapter 22: nAChRs in Non-Neuronal Tissues

While nicotinic acetylcholine receptors (nAChRs) are best known for their roles in the nervous system, recent research has illuminated their significant functions in non-neuronal tissues as well. The discovery that nAChRs are expressed in various peripheral tissues and organs has opened up exciting avenues for understanding their broader physiological roles. This chapter explores the functions of nAChRs in non-neuronal tissues, with a focus on immune cells, epithelial and endothelial cells, and their contribution to organ functions beyond the nervous system. We will also discuss the implications of these findings for novel therapeutic strategies targeting nAChRs in non-neuronal contexts.

1. Role of nAChRs in Immune Cells and Inflammation

nAChRs are expressed on a variety of immune cells, including macrophages, T cells, and neutrophils, where they play a pivotal role in regulating immune responses. These receptors influence the activation, differentiation, and function of immune cells, thereby impacting inflammation and immune system homeostasis.

Regulation of Immune Responses

- nAChRs modulate the immune system through both direct and indirect mechanisms. Activation of nAChRs on immune cells can affect cytokine release, cell signaling, and the migration of immune cells to sites of infection or injury. The activation of α7 nAChRs, in particular, has been shown to inhibit the release of pro-inflammatory cytokines, such as TNF-α, IL-1β, and IL-6, making this receptor subtype an important modulator of inflammation.
- Conversely, nAChR antagonism can lead to an enhanced immune response, which may have therapeutic potential in diseases characterized by immune suppression, such as cancer or certain viral infections.

Cholinergic Anti-Inflammatory Pathway

The cholinergic anti-inflammatory pathway, mediated by nAChRs, represents a key mechanism through which the nervous system regulates peripheral inflammation. This pathway involves the vagus nerve, which releases acetylcholine (ACh) that activates α7 nAChRs on immune cells. The subsequent signaling cascade reduces the production of inflammatory cytokines and helps resolve inflammation. Disruption of this pathway has been implicated in chronic inflammatory diseases, including rheumatoid arthritis, inflammatory bowel disease, and sepsis.

Therapeutic Implications

Targeting nAChRs to modulate immune responses holds promise for treating autoimmune diseases, chronic inflammatory conditions, and sepsis. For example, selective α7 nAChR agonists are being investigated as potential anti-inflammatory agents. Conversely, blocking nAChR signaling in certain cancers could enhance immune cell activity and tumor immune surveillance.

2. nAChRs in Epithelial and Endothelial Cells

Beyond their role in the immune system, nAChRs are also expressed in epithelial and endothelial cells, where they contribute to the regulation of tissue homeostasis and organ function.

Epithelial Cells

- nAChRs on epithelial cells have been shown to play a role in regulating mucosal secretions, wound healing, and epithelial integrity. For instance, in the respiratory system, nAChRs contribute to the regulation of airway smooth muscle tone and mucus production, processes that are important in diseases like asthma and chronic obstructive pulmonary disease (COPD). Similarly, in the gastrointestinal tract, nAChRs modulate gastrointestinal motility and mucus secretion.
- In skin and other epithelial tissues, nAChRs also participate in wound healing processes by influencing cell proliferation, migration, and differentiation. This suggests that nAChR modulation could be explored for therapeutic strategies in tissue repair and regeneration.

Endothelial Cells

- Endothelial cells, which line the blood vessels, are another key target for nAChR signaling. nAChR activation in endothelial cells has been implicated in the regulation of vascular tone and blood pressure. Through the release of nitric oxide (NO), nAChRs promote vasodilation and improve blood flow. This mechanism has important implications for cardiovascular health, especially in conditions like hypertension and atherosclerosis.
- Furthermore, nAChRs influence endothelial cell permeability and the formation of new blood vessels (angiogenesis), which is crucial for tissue regeneration and repair. These effects make nAChRs attractive targets for diseases that involve endothelial dysfunction, such as ischemic heart disease or stroke.

3. Contribution of nAChRs to Organ Function Beyond the Nervous System

While nAChRs are most commonly associated with neural signaling, their presence and activity in non-neuronal tissues suggest that they contribute significantly to the physiological functions of multiple organs.

The Cardiovascular System

In addition to endothelial cells, nAChRs play a role in the regulation of heart rate and contractility. In the heart, nAChRs are involved in modulating autonomic functions through parasympathetic signaling. The activation of muscarinic receptors, combined with nAChRs in the heart, contributes to the parasympathetic control of heart rate. This cholinergic control is crucial for maintaining normal cardiovascular function, and disruptions in this system can lead to arrhythmias and other cardiac disorders.

The Respiratory System

nAChRs also influence the function of the respiratory system, particularly in the regulation of smooth muscle contraction in the airways. In diseases like asthma and COPD, excessive bronchoconstriction is a hallmark, and nAChRs in airway smooth muscle cells contribute to this process. Modulating nAChRs in the lungs may help in the development of treatments for chronic respiratory diseases, reducing airway hyperreactivity and inflammation.

The Digestive System

Within the gastrointestinal system, nAChRs modulate motility, secretion, and inflammation. They influence the contraction of smooth muscles in the digestive tract and regulate peristalsis. Dysfunction in this system can lead to disorders such as irritable bowel syndrome (IBS) or gastroparesis. Targeting nAChRs could offer novel treatment strategies for improving gastrointestinal function and addressing motility disorders.

The Kidney and Renal System

nAChRs are also expressed in renal tissues, where they may affect glomerular filtration and fluid balance. Activation of nAChRs in kidney cells can influence sodium reabsorption, and thus, blood pressure regulation. This suggests that nAChRs may play a role in maintaining kidney function and fluid homeostasis, which could be of importance in conditions like hypertension and kidney disease.

4. Implications for Non-Neuronal Therapies

Understanding the roles of nAChRs in non-neuronal tissues has broad therapeutic implications. By modulating nAChR activity, it is possible to influence not just neuronal function but also immune responses, tissue repair, and organ function across a range of bodily systems.

Therapeutic Modulation of Non-Neuronal nAChRs

- Targeting nAChRs in non-neuronal tissues could offer novel treatments for diseases like cancer, chronic inflammation, cardiovascular diseases, and respiratory conditions. For example, selective modulation of α7 nAChRs in immune cells could help treat inflammatory diseases, while modulation of endothelial nAChRs could improve cardiovascular health.
- In cancer, nAChRs on tumor cells and the tumor microenvironment may contribute to tumor growth, metastasis, and immune evasion. Targeting these receptors could help in designing therapies that restrict tumor progression and enhance immune responses.

Potential for Personalized Medicine

As our understanding of nAChRs in non-neuronal tissues deepens, it will become increasingly possible to tailor therapies to individual patients. By identifying specific patterns of nAChR expression and activity in non-neuronal tissues, clinicians could optimize drug treatments and improve therapeutic outcomes for a wide array of diseases.

Challenges and Future Directions

- Despite the promise of nAChR-targeted therapies for non-neuronal tissues, several challenges remain. One major hurdle is the specificity of receptor modulation—ensuring that drugs selectively target nAChRs in specific tissues without causing adverse effects in others. Moreover, the complexity of nAChR signaling pathways in different tissues necessitates a nuanced approach to therapy development.
- Future research should focus on identifying novel ligands that selectively modulate nAChRs in non-neuronal tissues and exploring their potential for treating diseases that currently lack effective therapies.

Conclusion

nAChRs are integral to a wide range of physiological processes beyond their classic role in the nervous system. From modulating immune responses and inflammation to regulating vascular tone and organ function, these receptors contribute to the homeostasis of multiple organ systems. The therapeutic potential of targeting nAChRs in non-neuronal tissues is vast, offering promising strategies for treating diseases ranging from chronic inflammatory conditions to cancer and cardiovascular disorders. As our understanding of nAChR biology in these tissues continues to evolve, it is likely that we will see an expansion of nAChR-based therapies that offer new hope for patients with a broad spectrum of diseases.

Chapter 23: Environmental and Lifestyle Factors Affecting nAChRs

Nicotinic acetylcholine receptors (nAChRs) are integral to a broad range of physiological functions across the nervous and non-neuronal systems. However, their activity and expression can be influenced by various environmental and lifestyle factors. Understanding how diet, exercise, stress, environmental toxins, and smoking impact nAChRs is crucial not only for the development of therapeutic strategies but also for public health initiatives aimed at mitigating the effects of these factors. This chapter explores how external influences modulate nAChR expression and function, and their implications for health and disease.

1. The Effects of Diet, Exercise, and Stress on nAChR Expression

Diet and Nutrition

Diet plays a significant role in regulating the expression and function of nAChRs, particularly through its influence on the availability of neurotransmitters and the presence of dietary compounds that affect receptor activity. Several key nutritional factors can modulate nAChR expression and neurotransmission:

- **Choline and Acetylcholine Synthesis**: Choline, a precursor to acetylcholine, is crucial for maintaining healthy cholinergic signaling, including through nAChRs. Deficiencies in choline have been associated with cognitive decline, highlighting the importance of dietary choline (found in eggs, meat, and fish) in supporting nAChR function.

- **Omega-3 Fatty Acids**: Studies have shown that omega-3 fatty acids, particularly those from fish oil, can influence the expression of nAChRs. Omega-3 fatty acids are thought to enhance cholinergic signaling, potentially improving cognitive function and reducing the risk of neurodegenerative diseases such as Alzheimer's and Parkinson's.

- **Polyphenols and Antioxidants**: Polyphenolic compounds found in fruits, vegetables, and tea (e.g., curcumin and resveratrol) have been shown to affect nAChR expression and function. These compounds may help protect nAChRs from oxidative damage, thus supporting brain health and overall neurological function.

Exercise

Physical activity has a profound effect on brain health and can alter the expression of nAChRs. Exercise has been found to:

- **Increase nAChR Density**: Regular exercise, particularly aerobic exercise, has been shown to increase the density of nAChRs in the brain. This increase in receptor expression could enhance cholinergic neurotransmission, potentially improving cognitive function, learning, and memory.
- **Neuroplasticity**: Exercise-induced changes in nAChR expression are also thought to be linked to neuroplasticity—the brain's ability to form and reorganize synaptic connections. This process is essential for learning and adapting to new experiences.
- **Stress Reduction**: Physical activity reduces stress hormones like cortisol and increases endorphins, contributing to a balanced neurochemical environment that can support nAChR function. Chronic stress, on the other hand, can lead to decreased nAChR expression, particularly in the hippocampus, a region critical for memory and learning.

Stress and Hormonal Influences

Stress, whether acute or chronic, can have significant effects on the expression and functioning of nAChRs. During periods of stress, the body produces various hormones, such as cortisol and adrenaline, which influence receptor function:

- **Chronic Stress and nAChRs**: Prolonged stress has been associated with a decrease in nAChR density in the brain. This reduction may impair cognitive function, increase susceptibility to mood disorders like depression, and potentially contribute to neurodegenerative diseases. The mechanisms underlying these changes are complex but are thought to involve dysregulation of acetylcholine release and receptor desensitization.
- **Acute Stress and nAChRs**: On the other hand, short-term, acute stress might enhance cholinergic transmission in certain regions of the brain, improving focus and alertness. However, prolonged activation of the stress response system can lead to detrimental effects on nAChR function, highlighting the need for balance in stress management.

2. Environmental Toxins and Their Impact on nAChR Function

Environmental toxins, including pollutants, chemicals, and heavy metals, can significantly impact nAChR function. These compounds may interfere with receptor activity, leading to adverse effects on the nervous system and overall health.

Pesticides and Organophosphates

Organophosphates, commonly used in agricultural pesticides, are known to inhibit acetylcholinesterase, the enzyme responsible for breaking down acetylcholine in the synaptic cleft. The inhibition of this enzyme leads to an accumulation of acetylcholine, which can overstimulate nAChRs, potentially causing toxic effects on the nervous system, including:

- **Neurodegeneration**: Chronic exposure to organophosphates has been linked to neurodegenerative diseases, such as Alzheimer's disease, due to the dysregulation of acetylcholine systems and nAChRs.
- **Cognitive Impairment**: Long-term exposure can lead to deficits in learning, memory, and motor function, which are often associated with nAChR dysfunction. Organophosphates may also impair neuroplasticity by altering cholinergic signaling pathways.

Heavy Metals (Lead, Mercury, Cadmium)

Exposure to heavy metals, such as lead, mercury, and cadmium, can affect the functioning of nAChRs. These metals can bind to the receptors or interfere with the cellular signaling pathways that regulate their activity:

- **Neurotoxic Effects**: Lead exposure, for instance, has been shown to impair nAChR function by altering receptor expression and reducing cholinergic transmission. This can contribute to cognitive deficits and developmental delays in children.
- **Oxidative Stress**: Heavy metals can induce oxidative stress, which damages neurons and disrupts nAChR signaling. This damage contributes to neurodegeneration and can exacerbate conditions such as Alzheimer's disease and Parkinson's disease.

Air Pollution

Airborne pollutants, including fine particulate matter (PM2.5), ozone, and nitrogen dioxide, have been shown to affect brain health by influencing the cholinergic system and nAChR expression. Prolonged exposure to air pollution has been associated with:

- **Cognitive Decline**: Research suggests that pollutants may accelerate cognitive decline by impairing cholinergic signaling in the brain, thereby contributing to the onset of conditions like Alzheimer's disease and other forms of dementia.
- **Neuroinflammation**: Air pollution can also increase neuroinflammation, which affects the function of nAChRs and may contribute to the development of psychiatric disorders, such as depression and anxiety.

3. The Role of Smoking and Other Lifestyle Factors in nAChR-Related Diseases

Smoking and nAChRs

Cigarette smoking is one of the most well-established lifestyle factors affecting nAChR function. Nicotine, the primary psychoactive component of tobacco, exerts its effects by binding to nAChRs in the brain and other tissues, leading to both short-term and long-term changes in receptor function:

- **Acute Effects**: Nicotine's binding to nAChRs stimulates the release of dopamine, which produces feelings of pleasure and reward. This reinforces the habit of smoking and contributes to nicotine addiction.
- **Chronic Effects**: Long-term smoking leads to an upregulation of nAChRs, particularly in the brain, as the body compensates for the desensitization caused by chronic nicotine exposure. This upregulation is associated with the development of nicotine dependence and tolerance, as well as the heightened risk of diseases such as lung cancer, cardiovascular disease, and chronic obstructive pulmonary disease (COPD).
- **Neuroplasticity and Cognitive Function**: Smoking can also alter neuroplasticity by modulating nAChR expression in brain regions associated with learning and memory, such as the hippocampus. This may contribute to cognitive deficits in smokers and increase the risk of neurodegenerative diseases later in life.

Other Lifestyle Factors

In addition to smoking, other lifestyle factors such as alcohol consumption, diet, and sedentary behavior can influence nAChR function:

- **Alcohol**: Chronic alcohol consumption has been shown to alter nAChR expression, particularly in the brain regions involved in mood regulation and cognitive function. Alcohol-induced changes in nAChR activity are thought to contribute to the development of alcohol use disorders, as well as neurocognitive deficits.
- **Sedentary Lifestyle**: Lack of physical activity can exacerbate the negative effects of environmental toxins and poor dietary habits on nAChR function. Sedentary behavior is linked to an increased risk of cognitive decline and neurodegenerative diseases, partially due to its detrimental impact on cholinergic signaling and nAChR expression.

4. Public Health Perspectives on Nicotine Use and nAChR Modulation

The widespread use of nicotine, particularly through smoking, has significant public health implications. Understanding the effects of nicotine and other lifestyle factors on nAChR function is essential for developing effective public health policies and interventions aimed at reducing the harm caused by smoking and other nicotine-related behaviors. Strategies to combat nicotine addiction and its impact on nAChRs include:

- **Nicotine Replacement Therapies (NRTs)**: NRTs, such as nicotine patches, gums, and lozenges, are designed to help individuals quit smoking by providing a controlled, lower dose of nicotine to reduce withdrawal symptoms. These therapies work by modulating nAChR activity in a way that helps ease the transition away from smoking.
- **Smoking Cessation Programs**: Public health campaigns that promote smoking cessation and educate individuals about the risks of smoking can reduce the incidence of nicotine-related diseases, such as lung cancer and cardiovascular disease, which are linked to changes in nAChR function.

Chapter 24: Interdisciplinary Approaches to nAChR Research

Nicotinic acetylcholine receptors (nAChRs) are not only critical to the functioning of the nervous system but also play key roles in various non-neuronal tissues. As a result, research into nAChRs requires a collaborative, interdisciplinary approach that brings together knowledge from pharmacology, neuroscience, genetics, bioinformatics, and systems biology. Advances in our understanding of nAChRs—ranging from their molecular structure and function to their involvement in diseases—are dependent on the combined expertise of multiple scientific disciplines. This chapter will explore the benefits of these interdisciplinary collaborations, how they drive innovative research techniques, and how integrating diverse scientific fields can uncover new therapeutic opportunities.

1. Collaborations Across Pharmacology, Neuroscience, and Genetics

The study of nAChRs inherently involves several areas of expertise. Each discipline offers unique insights into the functioning of nAChRs at various levels, from molecular to systemic. By collaborating, researchers can explore the complexity of nAChR biology more comprehensively.

- **Pharmacology and Drug Discovery**: Pharmacologists play a vital role in studying how different compounds, both synthetic and natural, affect nAChRs. They provide essential information on agonists, antagonists, and other modulatory compounds that can influence nAChR activity. Their expertise is crucial for the design and development of drugs targeting nAChRs, which may have applications in treating neurodegenerative diseases, addiction, and other conditions.
- **Neuroscience and Functional Insights**: Neuroscientists contribute knowledge about how nAChRs influence neuronal signaling, cognition, behavior, and neurological disorders. Their work helps identify the roles of specific nAChR subtypes in different brain regions and their involvement in processes like learning, memory, and motor function. By conducting both in vivo and in vitro experiments, they generate valuable data on how nAChRs mediate neurotransmission and plasticity.

- **Genetics and Molecular Mechanisms**: Geneticists focus on the identification and characterization of genes involved in the synthesis of nAChRs, their expression, and their regulation. They study the genetic mutations that lead to nAChR dysfunction in diseases such as congenital myasthenic syndromes and provide insights into how genetic variations influence the response to drugs and susceptibility to disorders. Furthermore, advances in genetic engineering, such as CRISPR technology, allow for the creation of animal models with targeted mutations in nAChR genes to better understand the impact of specific receptor subtypes.

Together, these disciplines offer a multidimensional approach to understanding nAChRs and developing interventions that target their dysfunction.

2. The Importance of Bioinformatics in Studying nAChRs

Bioinformatics is an emerging field that combines computer science, biology, and statistics to manage and analyze complex biological data. It plays an increasingly important role in the study of nAChRs, especially given the vast amount of data generated by genomic and proteomic studies.

- **Genomic Databases and Receptor Mapping**: Bioinformatics tools are essential for organizing and analyzing the genetic data related to nAChRs. Databases like GenBank, Protein Data Bank (PDB), and others store genetic sequences of nAChR subunits, allowing researchers to study the molecular structure and evolutionary relationships of different receptor types. By comparing the genetic makeup of nAChRs across species, scientists can gain insights into receptor evolution and functionality.

- **Structural Bioinformatics**: Understanding the three-dimensional structure of nAChRs is crucial for drug discovery and the development of therapies targeting these receptors. Computational modeling, such as molecular docking studies, helps predict how different compounds interact with nAChRs at the atomic level. Advances in cryo-electron microscopy and X-ray crystallography have provided detailed structural insights into nAChRs, facilitating the design of more selective and effective drugs.

- **Data Integration and Systems Biology**: Bioinformatics allows researchers to integrate large datasets from genomic, transcriptomic, proteomic, and clinical studies. This data integration is critical for systems biology, which seeks to understand how nAChRs function within the context of the entire biological system. Systems biology approaches can uncover how nAChRs interact with other signaling pathways, providing a more holistic understanding of their role in health and disease.

3. **Insights from Systems Biology in Understanding nAChR Functions**

Systems biology approaches, which involve studying the complex interactions within biological systems, are invaluable in understanding the broader physiological functions of nAChRs. Rather than studying nAChRs in isolation, systems biology places them in the context of entire networks of molecules, cells, and organs.

- **Network Models of nAChR Signaling**: nAChRs are involved in complex signaling pathways that interact with other neurotransmitter systems, including dopamine, serotonin, and glutamate. Systems biology helps create models that describe how nAChRs contribute to the regulation of these pathways, revealing how receptor activation can modulate behaviors like learning, attention, and addiction. By understanding these networks, researchers can better understand how disruptions in nAChR function contribute to psychiatric disorders such as schizophrenia, anxiety, and depression.
- **Pathophysiological Insights**: nAChRs are implicated in numerous diseases, from neurodegenerative conditions like Alzheimer's and Parkinson's to autoimmune diseases like myasthenia gravis. Systems biology enables the exploration of how nAChR dysfunction leads to disease at a systems level. By examining gene expression changes, protein interactions, and cellular responses in diseased states, systems biology helps identify new therapeutic targets and biomarkers.

- **Multi-Omic Approaches**: Systems biology often incorporates multi-omic data (genomics, transcriptomics, proteomics, metabolomics) to better understand nAChR function across different biological levels. This approach is essential for identifying new molecules that interact with nAChRs and for understanding how environmental, genetic, and lifestyle factors affect receptor function. For example, by examining how certain diets or toxins influence nAChR-related pathways, systems biology can reveal new insights into disease prevention and treatment strategies.

4. Integrating nAChR Research with Other Fields Like Immunology and Cancer Research

While the bulk of research on nAChRs has focused on their role in the nervous system, there is growing interest in understanding how they influence non-neuronal functions. The interdisciplinary approach extends to fields like immunology and cancer research, where nAChRs are increasingly recognized for their roles beyond neurotransmission.

- **Immunology and Inflammation**: nAChRs are present not only in neurons but also in immune cells such as macrophages, T cells, and B cells. Research into the role of nAChRs in immune cell signaling has opened up new avenues for treating autoimmune diseases and chronic inflammation. Activation of nAChRs in immune cells has been shown to modulate the release of pro-inflammatory cytokines, suggesting that nAChR agonists could serve as potential anti-inflammatory agents.

- **nAChRs and Cancer**: Cancer research has also begun to focus on the role of nAChRs in tumor progression. nAChRs are expressed in various cancer cells, where they can influence processes such as cell proliferation, migration, and survival. By integrating cancer research with nAChR studies, scientists are exploring new strategies for targeting nAChRs in cancer treatment. For example, nAChR antagonists could be used to inhibit tumor growth, while nAChR-based biomarkers may aid in the early detection of certain cancers.

- **Pharmacological Synergies**: The integration of nAChR research with immunology and cancer biology may lead to the development of new therapeutic strategies that combine nAChR modulators with immune modulators or chemotherapeutic agents. This approach could enhance treatment efficacy, particularly in diseases where inflammation or immune dysfunction is central to disease progression.

5. The Future of Interdisciplinary nAChR Research

As the field of nAChR research continues to evolve, interdisciplinary collaborations will become even more essential. Emerging technologies, such as artificial intelligence (AI), machine learning, and high-throughput screening, will play a pivotal role in advancing our understanding of nAChR biology. AI can analyze large datasets, predict new compounds that could interact with nAChRs, and help identify novel therapeutic targets. Machine learning algorithms can also be used to uncover hidden patterns in nAChR-related diseases and predict patient responses to treatments.

Collaborations across disciplines will also foster the development of personalized medicine approaches, where therapies are tailored to an individual's genetic and molecular profile. By combining pharmacology, genetics, bioinformatics, and systems biology, researchers will be able to develop targeted therapies that address the specific nAChR-related dysfunctions of individual patients, leading to more effective and less toxic treatments.

Conclusion

Interdisciplinary approaches are essential for advancing the field of nAChR research. By integrating knowledge from pharmacology, neuroscience, genetics, bioinformatics, systems biology, and other fields, researchers can uncover new insights into nAChR function, identify novel therapeutic targets, and develop more effective treatments for a variety of neurological and non-neurological diseases. As the understanding of nAChRs deepens and interdisciplinary collaborations strengthen, the potential for breakthroughs in medical science will expand, ultimately leading to more targeted and personalized therapies for patients worldwide.

Chapter 25: Conclusion and Future Directions

As we reach the conclusion of our exploration into the fascinating world of nicotinic acetylcholine receptors (nAChRs), it is clear that these receptors are not merely integral to the transmission of signals across synapses, but also central to a vast array of physiological, pharmacological, and therapeutic processes. Our understanding of nAChRs has evolved dramatically, from their initial discovery to the groundbreaking research of their involvement in addiction, neurodegenerative diseases, and even cancer. As we close this journey, we will recap the critical insights gained throughout the book and discuss the future directions in nAChR research and therapy development.

1. Key Takeaways from the Book

Throughout the chapters, we have delved into the molecular biology, biochemistry, and pharmacology of nAChRs, uncovering the remarkable complexity of these receptors and their diverse roles in the nervous system and beyond. Key points include:

- **Molecular and Structural Diversity**: nAChRs are composed of various subunits, and their functional diversity is reflected in the variety of receptor subtypes expressed across different tissues. This allows them to participate in a range of processes, from cognition and motor function to immune regulation and cancer progression.
- **Physiological Significance**: In the central nervous system, nAChRs are essential for cognitive functions such as learning and memory, and their dysfunction is implicated in neurodegenerative disorders like Alzheimer's and Parkinson's. In the peripheral nervous system, they are involved in neuromuscular transmission, influencing both motor function and pain modulation.
- **Pharmacological Potential**: The therapeutic implications of nAChR modulation are vast, encompassing treatments for neurological disorders, mental health conditions, and addiction. Despite significant advancements, challenges remain in the design of selective drugs that target specific nAChR subtypes without causing harmful side effects.
- **Environmental and Lifestyle Influences**: Environmental factors, such as diet, stress, and toxins, have a profound impact on nAChR function. These influences underscore the importance of lifestyle choices in maintaining receptor function and preventing disease, particularly with regard to smoking and its detrimental effects on health.

- **Interdisciplinary Synergies**: The study of nAChRs is inherently interdisciplinary, involving genetics, bioinformatics, pharmacology, neuroscience, and more. Such collaborations are crucial for advancing research, uncovering new therapeutic strategies, and improving patient outcomes.

2. The Potential for Future Breakthroughs in nAChR Research

Looking ahead, the field of nAChR research is poised for exciting breakthroughs. Several factors are contributing to the rapid acceleration of discoveries and expanding therapeutic possibilities:

- **Advancements in Technology**: Technologies such as CRISPR gene editing, optogenetics, and advanced imaging techniques are transforming how nAChRs are studied in both basic and clinical research. These tools enable precise manipulation of nAChRs and the ability to observe their behavior in real-time at the molecular and cellular levels. As these technologies improve, researchers will be able to uncover deeper insights into nAChR function and dysfunction.
- **Artificial Intelligence and Machine Learning**: AI and machine learning algorithms are increasingly being applied to analyze large datasets generated from genomic, proteomic, and clinical studies. These tools can help identify novel therapeutic targets, predict patient responses to drugs, and even accelerate the discovery of new nAChR modulators. AI has the potential to revolutionize personalized medicine, enabling treatments that are tailored to the genetic and molecular profiles of individual patients.
- **The Role of Systems Biology**: Systems biology approaches, which look at the holistic interactions between genes, proteins, and cells, are providing new insights into how nAChRs function within complex biological systems. By integrating multi-omic data (genomics, transcriptomics, proteomics), systems biology can reveal how nAChRs interact with other signaling pathways, providing a more comprehensive understanding of their role in health and disease.

- **New Drug Targets and Therapies**: As we continue to understand the specific roles of different nAChR subtypes in various tissues, there is a growing opportunity to develop drugs that selectively target these receptors. This precision medicine approach could result in therapies that are more effective and less prone to side effects. Additionally, new approaches to drug delivery, such as nanoparticle-based systems, may enhance the bioavailability and specificity of nAChR-targeted therapies.

3. How Understanding nAChRs Can Shape Medical Practices

The evolving understanding of nAChRs is already shaping the way medical practitioners approach a wide range of diseases and disorders. In the future, it is likely that nAChR-targeted therapies will become more integral to clinical practice across various domains:

- **Neurodegenerative Diseases**: Advances in nAChR research could lead to more effective treatments for conditions like Alzheimer's and Parkinson's. By targeting specific nAChR subtypes involved in cholinergic dysfunction, new therapies may improve cognition, motor function, and overall quality of life for patients with these diseases.

- **Addiction and Mental Health**: The role of nAChRs in addiction is well-established, particularly in nicotine dependence. However, emerging research also suggests that nAChRs may play a role in the abuse of other substances, such as alcohol and cocaine. Therapies that target these receptors could provide novel treatment options for addiction, reducing relapse and promoting long-term recovery. Additionally, understanding how nAChRs affect mood regulation could lead to more effective treatments for conditions like depression, anxiety, and schizophrenia.

- **Cancer Treatment**: The discovery that nAChRs are involved in tumor progression has opened up exciting new possibilities in cancer treatment. By targeting nAChR signaling pathways in cancer cells, researchers may be able to inhibit tumor growth and metastasis. Further research could also establish nAChRs as biomarkers for early cancer detection, potentially leading to more effective screening methods.

- **Pain Management and Inflammation**: As nAChRs are involved in the modulation of pain and inflammation, therapies that modulate these receptors could offer new treatments for chronic pain conditions and inflammatory diseases, such as arthritis and autoimmune disorders.

4. Final Thoughts on Advancing Therapeutic Strategies Involving nAChRs

While significant strides have been made in understanding the biology and pharmacology of nAChRs, the field is still in its early stages, and many challenges remain. Continued research will be critical in identifying the full spectrum of nAChR subtypes and understanding their complex roles in both physiological and pathological processes. The interdisciplinary approach outlined in earlier chapters will be key to advancing our knowledge and unlocking the therapeutic potential of nAChRs.

Future breakthroughs in nAChR research will depend on overcoming several hurdles, such as developing highly selective drugs that target specific receptor subtypes, understanding the long-term effects of nAChR modulation, and ensuring the safety and efficacy of new therapies in clinical settings. As technology continues to advance and collaborations across disciplines grow, the potential for nAChR-based therapies to revolutionize the treatment of neurological, psychiatric, and non-neurological diseases is immense.

In conclusion, mastering the intricacies of nicotinic acetylcholine receptors presents a promising frontier in medical science. Through ongoing research and technological innovations, we are on the cusp of discovering new therapies that could improve countless lives. As our understanding of nAChRs deepens, we move closer to a future where these powerful molecules are harnessed to treat and prevent a wide range of diseases, ultimately contributing to a healthier, more resilient world.

www.ingramcontent.com/pod-product-compliance
Lightning Source LLC
Chambersburg PA
CBHW082245220526
45469CB00009B/2882